Marek's Disease

Biology of Animal Infections

Series Editor
Paul-Pierre Pastoret, Institute for Animal Health, Compton Laboratory, UK
ISSN: 1572-4271

PUBLISHED VOLUMES

Marek's Disease: *An Evolving Problem*
Edited by Fred Davison and Venugopal Nair, Institute for Animal Health, Compton Laboratory, UK
ISBN: 0-12-088379-1

FUTURE VOLUMES

Rinderpest and Peste des Petits Ruminants: *Virus Plagues of Large and Small Ruminants*
Edited by Paul-Pierre Pastoret, Institute for Animal Health, Compton Laboratory, UK and Thomas Barrett, Institute for Animal Health, Pirbright Laboratory, UK
ISBN: 0-12-088385-6

Scrapie

Foot-and-Mouth Disease

Bluetongue

African Swine Fever

Marek's Disease
An Evolving Problem

Edited by Fred Davison and Venugopal Nair

ELSEVIER
ACADEMIC
PRESS

Amsterdam • Boston • Heidelberg • London • New York • Oxford
Paris • San Diego • San Francisco • Singapore • Sydney • Tokyo

Elsevier Academic Press
84 Theobald's Road, London WC1X 8RR, UK
http://www.elsevier.com

Elsevier Academic Press
525 B Street, Suite 1900, San Diego, California 92101-4495, USA
http://www.elsevier.com

British Library Cataloguing in Publication Data
A catalogue record for this book is available from the British Library

Library of Congress Catalog Number: 2004104170

ISBN 0-12-088379-1
ISSN 1572-4271

The front cover shows high power images of Marek's disease herpesvirus (RB1B)-infected chicken embryo cells stained with fluorescent-conjugated antibodies against the viral proteins. The left image shows cells stained for VP22 (green), the middle one ICP4 (red) and the right image VP22 (green) with VP5 (red). These images were kindly provided by Dr Jean-François Vautherot, Centre de Recherches de Tours, INRA, France.

Transferred to Digital Print 2007
Printed and bound by CPI Antony Rowe, Eastbourne

Dedication

Nat Bumstead (1952–2004)

This book is dedicated to the memory of Nat Bumstead. As well as being a creative, gifted and highly-respected scientist, he was our valued colleague and friend, who showed great courage, optimism and zest for life. His contributions to the fields of avian genetics and Marek's disease research are greatly appreciated. He will be sorely missed.

Dedication

Jan Brunstead (1955–2004)

This book is dedicated to the memory of Jan Brunstead. As well as being a creative, gifted and highly respected scientist, he was our valued colleague and friend, who showed great courage, optimism and zest for life. His contributions to the fields of avian genome and Marek's disease research are greatly appreciated. He will be sorely missed.

Contents

Colour plates appear between pages 116 and 117

Contributors

Susan J. Baigent
Institute for Animal Health, Compton Laboratory, Newbury, Berkshire, UK

Peter M. Biggs
Willows, London Road, St Ives, Cambridgeshire, UK

Michel Bublot
Virology Unit, Discovery Research, Merial, Lyon, France

Nat Bumstead
Formerly Institute for Animal Health, Compton Laboratory, Newbury, Berkshire, UK. (Nat Bumstead died on 20th April 2004 before publication of this book.)

Shane C. Burgess
Department of Basic Sciences, College of Veterinary Medicine, Mississippi State University, Mississippi, USA

Fred Davison
Institute for Animal Health, Compton Laboratory, Newbury, Berkshire, UK

Frank Fehler
Lohmann Animal Health GmbH & Co., Cuxhaven, Germany

Isabel M. Gimeno
Avian Disease and Oncology Laboratory, East Lansing, Michigan, USA

Jim Kaufman
Institute for Animal Health, Compton Laboratory, Newbury, Berkshire, UK

Pete Kaiser
Institute for Animal Health, Compton Laboratory, Newbury, Berkshire, UK

Hsing-Jien Kung
UC Davis Cancer Center, UCDMC, Sacramento, California, USA

Chris Morrow
Aviagen Ltd, Newbridge, Midlothian, Scotland

Venugopal Nair
Institute for Animal Health, Compton Laboratory, Newbury, Berkshire, UK

Klaus Osterrieder
Department of Microbiology and Immunology, College of Veterinary
Medicine, Cornell University, Ithaca, New York, USA

Paul-Pierre Pastoret
Institute for Animal Health, Compton Laboratory, Newbury, Berkshire, UK

Laurence N. Payne
Formerly Institute for Animal Health, Compton Laboratory, Newbury,
Berkshire, UK

Karel A. Schat
Department of Microbiology and Immunology, College of Veterinary
Medicine, Cornell University, Ithaca, New York, USA

Jagdev Sharma
Department of Veterinary and Biomedical Sciences, College of Veterinary
Medicine, University of Minnesota, St Paul, Minnesota, USA

Jean-François Vautherot
Centre de Recherches de Tours, INRA, Nouzilly, France

Vladimír Zelník
Lohmann Animal Health GmbH & Co., Cuxhaven, Germany

Series Introduction: Biology of Animal Infections

The Institute for Animal Health (IAH) in the United Kingdom is one of the eight Institutes sponsored by the Biotechnology and Biological Sciences Research Council (BBSRC).

The Institute was formed following the merger of four previously independent institutes, including the Houghton Poultry Research Station, situated near Cambridge, which specialized in infectious and parasitic diseases of poultry. It became the Houghton Laboratory, and was moved to Compton in 1992. As a result, the Institute has three laboratories, located at Compton (Berkshire), Pirbright (Surrey), and Edinburgh (Scotland).

The IAH is a world-leading Institute for multidisciplinary research on infectious diseases of livestock. Its remit is to understand the processes of infectious diseases and from that knowledge improve the efficiency and sustainability of livestock farming, enhance animal health and welfare, safeguard the safety and supply of food, and protect the environment. The Institute also undertakes disease surveillance for the Department of the Environment, Food and Rural Affairs (DEFRA) in the United Kingdom, and reference laboratory services for DEFRA, the Office International des Epizooties – World Organisation for Animal Health (OIE), the Food and Agriculture Organisation (FAO), and the European Union (EU).

The laboratory at Compton concentrates on enzootic diseases, including some work on transmissible spongiform encephalopathies (TSEs). The laboratory at Pirbright specializes in exotic viral diseases, and the Neuropathogenesis Unit at Edinburgh works entirely on TSEs.

The IAH works on most of the list A diseases of the Office International des Epizooties, including foot-and-mouth disease, rinderpest, peste des petits ruminants, African horse sickness, bluetongue, African swine fever, classical swine fever, swine vesicular disease, lumpy skin disease, and sheep and goat pox. As far as enzootic diseases are concerned, the IAH works on some of the most important ones, including scrapie and bovine spongiform encephalopathy (as well as the human form new variant Creutzfeldt Jakob disease), Marek's disease, bovine viral diarrhoea, bovine respiratory syncytial infection, bovine tuberculosis, avian infectious bronchitis, avian poxvirus infections, avian leukosis and avian coccidiosis. It also works on bacteria responsible for food poisoning in humans, such as Salmonella, Campylobacter and *Escherichia coli*. All this work is supported by a strong expertise in immunology, genetics, molecular biology, microbiology and epidemiology (including mathematical modelling).

Recently the IAH has decided to share with the scientific community its longstanding expertise and experience acquired with several of these diseases, in the form of a series of monographs. Each of the monographs is devoted to a single disease and developed according to a similar outline, including historical background, disease description, pathology, aetiology, molecular biology of the infectious agent, immune responses of the host against infection, epidemiology and transmission, followed by more practical approaches such as prophylaxis and vaccination. The monographs are also intended to give access to all of the most relevant literature related to the disease in order to constitute a practical and invaluable source of information. They are aimed at readers in university veterinary departments and specialists in veterinary research institutes worldwide. They should also be invaluable to final year undergraduate students pursuing immunological, microbiological, virological, epidemiological and veterinary or medical studies.

The title of the series is *Biology of Animal Infections*. The reasons for using the term 'infection' instead of 'disease' are manifold. Infection of an animal does not always result in a disease; the outcome of an infection depends on many factors, including the genetics of the host and the longstanding evolutionary relationship between the host and its parasite. Many infections can result in persistence of the so-called pathogens in the host, leading to sub-clinical carriers of the infectious agent, which can be of crucial importance in the transmission and epidemiology of the infection.

The term 'biology' refers to the fact that the focus of the monographs is the biology – the mechanism – of the infection rather than a more descriptive approach to the disease, though this is equally important.

The editors of the monographs are convinced that studying infections (or diseases) in the actual (target) host is the best way to gain an understanding of health. The name 'Institute for Animal Health' is a true illustration of this concept.

The first monograph is devoted to Marek's disease. The reason for this choice is not only the fundamental scientific significance of this disease, which crosses a number of biological disciplines related to animal health, but also because the Institute is organizing the 7th International Marek's Disease Symposium at the University of Oxford in July 2004.

This first monograph will very quickly be followed by another on rinderpest, the first animal infection to be eradicated after smallpox. Other monographs already scheduled will deal with bluetongue, scrapie, foot-and-mouth disease, African swine fever etc. ...

Paul-Pierre Pastoret, Series Editor

Preface

Marek's disease was first described by József Marek in 1907 although the publication (*Dt. Tierärztl. Wschr.* 15, 417–421), which described the disease as a polyneuritis, did not stimulate interest for over a decade. This was probably because it was not recognized as a common or economically important disease at the time. The level of interest in the disease began to change in the 1920s, when a number of significant publications appeared, including a description of the disease in a number of states of the USA and the first indication that visceral lymphoid tumours were a manifestation of the disease.

As the poultry industry developed, the disease became more important, and the number of publications in the 1930s and 1940s increased dramatically. Many of these publications were the result of studies in the USA, but a significant number came from Europe, particularly Germany and Britain, reflecting the increasing economic importance of the disease to the poultry industries of those countries. It is interesting that although there are indications in many of the publications that the identity of the disease and its infectious nature was understood, none convincingly defined the disease or demonstrated its transmissibility. In the 1950s the poultry industry was developing rapidly, and Marek's disease took on a new form that was to have devastating consequences for the poultry industry around the world. This form was first noted in the USA as an increase in the incidence of the disease with a high proportion of affected chickens having visceral lymphoid tumours. By the mid-1960s this form was described as an epizootic disease in the Eastern Seaboard of the USA, and during the 1960s it had spread to most countries with well-developed poultry industries.

The 1960s was a period of intense research on Marek's disease and, most importantly, a period when funds were made available to provide isolation facilities and strains of chicken susceptible to the disease. Over a period of less that ten years the disease had been convincingly transmitted, shown to be contagious and to be caused by a herpesvirus, and vaccines had been developed that were to be the saviour of the poultry industry throughout the world. Over the years since then, and in the face of vaccination, the virus has mutated to ever-increasingly virulent forms, requiring new strategies for the control and prevention of the disease. For this reason, and for the comparative medical interest in a disease caused by a herpesvirus that encompasses both inflammatory and neoplastic components in its pathogenesis, research has remained important and vibrant to this day. The results of this research are reviewed in this monograph.

Peter M. Biggs

Acknowledgements

Elsevier has enthusiastically agreed to publish the series of monographs, and we are very grateful to them.

We would like to thank not only the Elsevier editors, but also Elaine Collier, Tracey Duncombe, Fred Davison, Venugopal Nair, Thomas Barrett, Steve Archibald, Mick Gill and all the members of the Institute contributing to this series.

Gathering this wealth of information would not have been possible without the commitment, dedication and generous participation of a large number of contributors from all over the world. We are greatly indebted to them for the considerable amount of work and their willingness to set aside other priorities to meet a very tight schedule for this project.

Paul-Pierre Pastoret

Abbreviations

ADCC	Antibody-dependent cell cytotoxicity
B-	Bursal-derived (cells)
BAC	Bacterial artificial chromosome
bp	Base pair
CD	Cluster of differentiation (of cellular antigenic marker)
CEC	Chick embryo cells
CIAV	Chicken infectious anaemia virus
CNS	Central nervous system
DNA	Deoxyribonucleic acid
E	Early or β (gene)
EDS-67	Egg drop syndrome-67
EHV	Equine herpesvirus
EST	Expressed sequence tagged
FFE	Feather-follicle epithelium
g	Glycoprotein (e.g. glycoprotein B)
GaHV	Gallid-herpesvirus
GC	Guanine and cytosine
GFP	Green fluorescent protein
GH	Growth hormone
HHV-8	Human herpesvirus 8
HSV	Herpes simplex virus
HTLV	Human T-cell leukaemia virus
HVS	*Herpesvirus saimiri*
HVT	Herpesvirus of turkeys
IB	Infectious bronchitis
IBDV	Infectious bursal disease virus
IBV	Infectious bronchitis virus
ICP	Infected cell protein
IE	Immediate early or α (gene)
IL-1	Interleukin-1

IRL	Internal repeat long
IRS	Internal repeat short
ISH	*In situ* hybridization
L	Late or γ (gene)
LAT	Latency-associated transcripts
LL	Lymphoid leukosis
mAbs	Monoclonal antibodies
MD	Marek's disease
MDCC	MDV-tranformed chicken cell-line
MDV	Marek's disease virus
MDV-1	MDV serotype 1
MDV-2	MDV serotype 2
MDV-3	MDV serotype 3, i.e. HVT
meq	MDV EcoRI-Q
MHC	Major histocompatibility complex
MSRs	MDV small RNAs
MSV	Master seed virus
ND	Newcastle disease
NKC	Natural killer complex
NOR	Nucleolar organizing region
ORFs	Open reading frames
PBL	Peripheral blood leukocytes
PCR	Polymerase chain reaction
PND	Persistent neurological disease
pp38	Phosphoprotein 38
RDA	Representational difference analysis
REV	Reticuloendotheliosis virus
RSV	Rous sarcoma virus
SAR	Small antisense RNA
ssDNA	Single-stranded DNA
T-	Thymus-derived (cells)
TAP	Transporters associated with processing
TCR	T cell receptor
T_H	T helper (cell)
α-TIF	α-transinducing factor
TP	Transient paralysis
TRL	Terminal repeat long
TRS	Terminal repeat short
UL	Unique long region of MDV genome
US	Unique short region of MDV genome
vIL-8	Viral homologue of interleukin-8
vMDV	Virulent Marek's disease virus
vv+MDV	Very virulent plus Marek's disease virus
vvMDV	Very virulent Marek's disease virus
VZV	*Varicella zoster* virus

Introduction

1

PAUL-PIERRE PASTORET

Institute for Animal Health, Compton Laboratory, Newbury, Berkshire, UK

General background

Marek's disease (MD), first described by József Marek in 1907 (Marek, 1907) as fowl paralysis (polyneuritis), was later shown to be associated with lymphoid tumours. The clinical disease has similarities to avian leukosis, and it was about 40 years before the two conditions were differentiated and the causative agent, Marek's disease herpesvirus (MDV), was identified. The acute lymphoid form of MD became particularly serious with the expansion and intensification of poultry production in the 1960s. The introduction of vaccination in the 1970s was a major breakthrough, this being the first demonstration of effective and widespread use of vaccination to prevent a virus-induced cancer in any species. The problems of MD morbidity and mortality receded, but vaccine breaks began to be reported and increased virulence of challenge viruses occurred within ten years. The subsequent introduction of more aggressive vaccines and vaccine regimes has driven MDV to evolve to even greater virulence over the last 30 years.

MDV has some fascinating features. It has a genomic structure similar to herpes simplex virus (an α-herpesvirus), but infects lymphocytes and behaves biologically more like Epstein-Barr virus (a γ-herpesvirus). The MDV life cycle is complex. MDV infects and destroys the antibody-producing B lymphocytes before infecting T lymphocytes that make up the regulatory and killer populations of the immune system. One week after infection, MDV becomes latent in the T cells. The genetics of the host play a crucial role in the disease outcome. In a resistant host the virus continues to remain latent, whereas in a susceptible host MDV neoplastically-transforms regulatory T lymphocytes, causing tumours to form in many types of soft tissue (acute MD) and/or causing paralytic lesions in the nerves (classical MD).

MD is of great interest to geneticists and tumour immunologists alike, providing a natural animal model to study the molecular mechanisms of carcinogenesis and cancer regression, as well as the importance of genetics in determining disease outcome. The clinical manifestations of MD continue to change and MDV continues to be a threat to poultry production, despite the widespread and intensive use of vaccination, including vaccination of the late embryo – needless to say, a vaccination strategy devised to help control MD. Studies on MD caution us of the need for sustainable vaccination strategies in future. The problems associated with MD vaccination also provide useful lessons for other species, especially the human population.

Over the past 20 years, there have been two major publications specifically devoted to MD. The standard text, *Marek's Disease: Scientific Basis and Methods of Control*, edited by L. N. Payne and published in 1985, still provides a very valuable and wide-ranging description of the virus, the disease and the host responses. The more recent book *Marek's Disease*, edited by Kanji Hirai and published in 2000, provides substantial descriptions of the virus, the disease and host genetics, as well as covering the history and biology of MDV. Our new publication tries to bring the subject of MD up to date, taking account of very recent developments in the molecular biology of the virus and information on the responses of the host. It makes MD, MDV and MD vaccination a very central theme – trying to show what lessons can be learned about herpesviruses; the use of vaccination against neoplastic disease; the development of *in ovo* vaccination technology; changes in the virus genome that are linked with changes in pathology and oncogenesis; evidence that vaccines are driving the pathogen to increasing virulence; and the need to develop sustainable vaccination strategies in future etc. This monograph intends to provide a primary and invaluable source of major references on MD and the causative virus, making it a useful and up-to-date handbook.

It also shows the pioneering role played by the former Houghton Poultry Research Station, now part of the Institute for Animal Health (IAH), in the isolation and identification of the causative herpesvirus, as well as in the first application of the vaccination strategy for controlling MD. Alongside the IAH, a number of other research laboratories worldwide have also made invaluable contributions to the study and control of this major burden on the poultry industry. Among the many Institutions that deserve to be mentioned, contributions from the United States Department of Agriculture Regional Poultry Laboratory at East Lansing (now the USDA Avian Disease and Oncology Laboratory) and the Department of Avian and Aquatic Medicine, Cornell College of Veterinary Medicine, USA, have provided seminal information on the biology and control of MD.

József Marek (1868–1952) and the identification of the disease

József Marek was born on 18 March 1868 into a simple farming family in Vágszerdahely, Nyitra County, which today is Horna Streda in Slovakia.

On 1 September 1889 he registered as a student at the Royal Hungarian Veterinary School in Budapest, where he completed the course on 5 November

Figure 1.1   József Marek (1868–1952).

1892. He obtained the veterinary diploma with a rarely seen high mark. At that time the name of the institution was changed to become the Royal Hungarian Veterinary College. József Marek was the contemporary of another prominent Hungarian veterinarian, Aladar Aujeszky (1869–1933), the discoverer of Aujeszky's disease (pseudorabies) in 1902. In 1894 Marek passed the veterinary official examination and published his first paper entitled 'Case report of Basedow's Disease in a horse', which appeared in *Veterinarius* (Number 7). In the summer of 1895 he was assigned to direct the laboratory of the Royal Veterinary Institute in Budapest – Kôbánya – and was awarded a Doctorate of Philosophy (Summa Cum Laude) on 25 April 1898. On 31 January 1901, József Marek was appointed Ordinary Teacher of the Royal Veterinary College and Director of the Internal Medicine Department. He held this position until 1935. With this appointment new dimensions appeared in the development of veterinary internal medicine. Besides raising the levels of both practical and theoretical education, he also successfully unified the areas of diagnosis, therapy and research. As a scientist, Marek dealt with almost the whole area of animal medicine, as was often the case at that time.

He worked notably on the histopathology of classical swine fever, and opposed former ideas in order to prove that *dourine* is also caused by trypanosomes in Hungary.

Marek's disease, so named in his honour, was initially discovered and described by Marek in two publications in 1907. The first, on clinical diagnosis, was published in Budapest in 1902. The three volumes of *Veterinary Internal Medicine*, edited together with Ferenc Hutÿra, were published in 1904, and a second edition in two volumes in 1923/1924. This work helped to lay the foundations for the discipline of veterinary internal medicine based on scientific methodology.

The two volumes of *Spezielle Pathologie und Therapie der Haustiere* were published in 1905 by G. Fisher of Jena. The work was produced in eleven editions, in German, until 1959. Marek collaborated with Ferenc Hutÿra, Rezso Manninger and János Mócsy on several of the later editions. The importance of this work is shown by the fact that it was translated into English, Italian, Russian, Spanish, Turkish, Serbian, Polish and Chinese. A partial translation into Finnish was also available. Another book, *Lehrbuch der Klinischen Diagnostic der inneren Krankheiten des Haustiere*, edited in 1912, was republished four times.

Apart from writing popular articles that appeared mainly in farming magazines, József Marek summarized the results of his scientific findings and research in 154 papers. He died on 2 September 1952 at the age of 84, and his obituary was published in the *Magyar Állatorvosok Lapja*. His grave can be found in the Farkasrét Cemetery of Budapest.

Houghton Poultry Research Station and the isolation of the causative agent of Marek's disease

One of the major steps forward in the history of MD occurred some six decades after the first description of the disease by Marek, with the isolation of the causative herpesvirus, MDV, at the Houghton Poultry Research Station, Houghton, Huntingdon, Cambridgeshire (before it moved to Compton, within the IAH). Propagation of the causative agent in cell culture by Churchill and Biggs (1968) and independently in the USA by Nazerian *et al.* (1968) and Solomon *et al.* (1968) was a major breakthrough in MD research, and this was soon followed by transmission of the disease with the cell culture-propagated virus. Eventually transmission of MD was achieved using cell-free infectious herpesvirus particles isolated from the feather follicles of infected chickens by Calnek and colleagues at the Cornell College of Veterinary Medicine (Calnek *et al.*, 1970), providing definitive proof that dispelled any doubts: MDV was indeed the causative agent.

Within two years of the discovery of the causative herpesvirus, the virulent isolate HPRS-16 was attenuated by Churchill and his colleagues at Houghton and used as an effective vaccine (Churchill *et al.*, 1969). Thus MD became the first and

Figure 1.2 A view of Houghton Poultry Research Station and gardens in the spring sunshine (Circa, 1973).

most important disease model in which a neoplastic condition was successfully controlled by vaccination. Soon after this, scientists at the USDA-ADOL, East Lansing, showed that the naturally avirulent herpesvirus of turkeys (Kawamura *et al.*, 1969; Witter *et al.*, 1970) was able to provide good protection against MD (Okazaki *et al.*, 1970). In 1972, Rispens and his colleagues in the Netherlands described the use of the attenuated MDV strain CVI988 (also known as the Rispens strain), a highly immunogenic strain that is used widely today as the most effective vaccine. It is remarkable that, before the availability of vaccines, the economic impact of MD on the poultry industry was so great that chicken strains resistant to MD were being selected despite losing some important production traits. However, the availability of MD vaccines superseded this approach and paved the way for vaccination as an economically more sustainable way of controlling MD.

Today, three distinct avian herpesviruses that share common antigenic determinants are recognized: Marek's disease herpesvirus 1 (*Gallid herpesvirus* 2); Marek's disease herpesvirus 2 (*Gallid herpesvirus* 3); and turkey herpesvirus 1 (*Meleagrid herpesvirus* 1). A major step in the study of the molecular biology of MDV was initiated by Ross and his colleagues at Houghton in 1988, and they demonstrated that MDV and HVT DNA sequences have greater similarity to the varicella zoster virus (*Human herpesvirus* 3), an α-herpesvirus, than to the Epstein-Barr virus (EBV), a γ-herpesvirus, and that the genomes of MDV and HVT are co-linear with those of varicella zoster virus. These studies showed that although some biological aspects of MD resembled EBV infection in humans, virologically it could not be considered to be a true model for human EBV infection because MDV is clearly an α-herpesvirus.

The place of Marek's disease and Marek's disease virus in the biological sciences

As listed below, poultry have played a crucial role in the development of both immunology and vaccinology (Davison, 2003):

- Graft versus host responses and the key role of lymphocytes in adaptive immunity were first described by work on chicken embryos and chickens.
- The bursa of Fabricius, as demonstrated by Glick and collaborators in 1956, provided the first substantive evidence that there are two major lineages of lymphocytes.
- Gene conversion, the mechanism used by the chicken to produce its antibody repertoire, was first described in the chicken, and requires the unique environment of the bursa.
- The chicken's major histocompatibility complex (MHC) is minimal and compact. Uniquely, the chicken MHC is strongly associated with resistance to infectious diseases.
- The first attenuated vaccine was developed by Louis Pasteur against a chicken pathogen, fowl cholera.
- Vaccination of chick embryos on the eighteenth day of incubation has been proved to provide protection early after hatching.
- Evidence that widespread and intensive vaccination can lead to increased virulence with some pathogens, such as Marek's disease virus and infectious bursal disease virus (Gumboro disease), was first described with chicken populations.

Similarly, vaccination against MD is considered to be the first example of widespread use of a vaccine to effectively control a naturally occurring cancer agent. Although MD vaccine was primarily developed for protecting chickens, its importance extends beyond the field of animal health and it has contributed to our understanding of related human diseases and fundamental biology (Payne, 1985), and of comparative herpesvirology and oncology (Calnek, 1986).

MDV is a lymphotrophic herpesvirus, which, after an early cytolytic infection, induces T-cell lymphomas in the chicken its natural host. The lymphoma cells are latently infected, and the integration pattern of MDV DNA suggests a clonal nature of tumour formation; although further studies are needed to confirm this, MDV-transformed cell lines established *in vitro* maintain the integration pattern of primary lymphomas (Delecluse *et al.*, 1993). The complexity of MD is further demonstrated by the fact that a number of factors – viral, host and environmental – can influence the outcome of infection by MDV. Strains of MDV vary greatly in their oncogenic potential, ranging from non-oncogenic to highly oncogenic, but viruses not normally considered oncogenic can be so in the appropriate host environment. MD provides an excellent example of this, where the genotype of the chicken has a major influence on the outcome of infection. Extensive research at the IAH, East Lansing, Cornell and other laboratories has provided valuable insights into the mechanisms of genetic resistance to MD. One environmental factor that clearly plays a role in MD, as well as several other poultry diseases, is stress. However, further studies are required to obtain robust scientific evidence on how this influences the outcome of MDV infection.

Several other interesting, complex features of MD are described in this comprehensive monograph. Due to the specific nature of this infection, as well as the more usual chapters, new ones (such as molecular oncogenicity, genetic resistance and future strategies for controlling the disease) have been included.

Acknowledgements

Writing this introduction would not have been possible without the help of Dr Vöräs Károly and Professor Solti Lászlö, Dean of the Veterinary College of Budapest, where József Marek did his seminal work. The photograph of József Marek was kindly provided by Mrs Eva Dren who obtained it from Mr Iván Gábor. I thank them warmly.

I am also greatly indebted to Peter Biggs, Jim Payne, Fred Davison and Venugopal Nair for their invaluable help.

References

Calnek, B. W. (1986). *CRC Crit. Rev. Microbiol.*, **12**, 293–320.
Calnek, B. W., Adldinger, H. K. and Kahn, D. E. (1970). *Avian Dis.*, **14**, 219–233.
Churchill, A. E. and Biggs, P. M. (1968). *J. Natl Cancer Inst.*, **41**, 951–956.
Churchill, A. E., Payne, L. N. and Chubb, R. C. (1969). *Nature*, **221**, 744–747.
Davison, T. F. (2003). *Br. Poultry Sci.*, **44**, 6–21.
Delecluse, H. J., Schüller, S. and Hammerschmidt, W. (1993). *EMBO J.*, **12**, 3277–3286.
Hirai, K. (2000). *Marek's Disease*. Springer, Berlin.
Kawamura, H., King, D. J. Jnr. and Anderson, D. P. (1969). *Avian Dis.*, **13**, 853–863.
Marek, J. (1907). *Dtsch. Tierärztl. Wochenschr.*, **15**, 417–421.
Nazerian, K., Solomon, J. J., Witter, R. L. and Burmester, B. R. (1968). *Proc. Soc. Exp. Biol. Med.*, **127**, 177–182.
Okazaki, W., Purchase, H. G. and Burmester, B. R. (1970). *Avian Dis.*, **14**, 413–429.
Payne, L. N. (1985). *Marek's Disease: Scientific Basis and Methods of Control*. Martinus Nijhoff Publishing, Boston.
Rispens, B. H., Van Vloten, H., Mastenbroek, N. *et al.* (1972). *Avian Dis.*, **16**, 108–125.
Solomon, J. J., Witter, R. L., Nazerian, K. and Burmester, B. R. (1968). *Proc. Soc. Exp. Biol. Med.*, **127**, 173–177.
Witter, R. L., Nazerian, K., Purchase, H. G. and Burgoyne, G. H. (1970). *Am. J. Vet. Res.*, **31**, 525–538.

Marek's disease – long and difficult beginnings

2

PETER M. BIGGS

Willows, London Road, St Ives, Cambridgeshire, UK

Introduction

Marek's disease (MD) came into being as the name for a disease of poultry in 1960, following papers presented at the First World Veterinary Poultry Association Conference in Utrecht by Campbell (1961) and Biggs (1961). Both suggested a separation of what we now know as MD from lymphoid leukosis (LL). The latter author suggested the name 'Marek's disease', because the previously used pathological names had led to confusion and the exact pathological nature of MD was uncertain at that time. This suggestion was formalized by the Conference adopting a resolution classifying the 'leucosis complex and fowl paralysis' (Biggs, 1962). This was the culmination of many years' confusion between these diseases, resulting in difficulties in the interpretation of research results.

Early days

The story began in the first decade of the last century when MD was first described (Marek, 1907), and the early classic study of the leukoses was published by Ellerman and Bang in 1908 (see Ellermann, 1921). The early descriptions of MD (Marek, 1907; Kaup, 1921; van der Walle and Winkler-Junius, 1924) suggested that it was a condition affecting only the nervous system, which was variously described as *polyneuritis*, *paralysis of the domestic fowl*, and *neuromyelitis gallinarum*. At that time it was clearly a different condition from the leukoses (Ellerman, 1921). The difficulty began to emerge with the publication of the excellent work of Pappenheimer and his colleagues (Pappenheimer *et al.*, 1926, 1929a, 1929b). These authors described lymphomas in six of the sixty field cases of paralysis they examined. They considered that these formed part of the disease syndrome because the incidence was higher than in normal chickens but also, and more importantly, because the cytological composition of the visceral lymphomas was similar to the lymphoid infiltrations seen in nervous tissue. Because of this finding they suggested the name *neuro-lymphomatosis gallinarum* for the disease. It is of interest that they also first used the term visceral lymphomatosis to describe the visceral tumours. The presence of lymphoid tumours in this disease, now known as Marek's disease, and in the condition described as lymphatic leukosis by Ellerman (1921) and now known as lymphoid leukosis, led to confusion over recognition of the two diseases because of the difficulties at the time in differential diagnosis.

The difficulty in the differential diagnosis between lymphoid tumours of MD and LL led many to view these diseases as a single entity. These included those who considered the whole complex of the leukoses and fowl paralysis to be variants of a single disease. Others confused LL with MD, resulting from the use of the term lymphomatosis. Still others were working with one or the other disease using generic terms such as lymphomatosis or leukosis to describe the condition they were working with. They were unaware that there were two diseases, and therefore did not realize with which they were working.

Terminology and classification

To develop some order and uniformity in the terminology and classification used in publications and enable better interpretations of reported data, a committee was set up, chaired by Erwin Jungherr, to propose a terminology and classification that, it was hoped, could be universally used. The committee proposed the term *lymphomatosis* for fowl paralysis (MD), dividing it into neural, ocular and visceral lymphomatosis (Jungherr, 1941). Although it was made clear that this was a pathological classification and had no aetiological basis, it inadvertently implied that LL was the same disease as MD because the lymphoid tumours of both diseases were included under the term visceral lymphomatosis.

A number of workers considered MD to be a distinct entity and separate from the leukoses, notably Fritzsche (1939) and Campbell (1945, 1956). An attempt was made by Chubb and Gordon (1957) to accommodate this view by suggesting a classification and terminology that separated the two diseases by restricting the use of the term lymphomatosis to MD, subdividing it into neural, ocular and visceral lymphomatosis, and suggesting LL for the second disease, but unfortunately this was not widely adopted. The Jungherr classification modified by Cottrall (1952), however, was leading to confusion over the two diseases and difficulty in interpretation of published studies as it was uncertain which disease was the subject of the work. The outcome was that it was impossible to interpret the data reported from the many laboratories investigating this group of diseases and for these data to be applied advantageously in the field.

A prominent example included the work of the Regional Poultry Laboratory at East Lansing and that of Hutt and Cole at Cornell. Both groups thought they were working on the same disease, but could not understand why their results were not consistent. A prime illustration of the problem was the claim by Hutt and Cole (Cole and Hutt, 1951; Hutt, 1951) that lymphomatosis was not egg-transmitted, whereas the Regional Poultry Laboratory group claimed it was (Cottrall *et al.*, 1954), with both providing evidence for their claims. We now know that the Regional Poultry Laboratory was working on LL and Cornell on MD. Waters (1954), at the Regional Poultry Laboratory, suspected that they were working on different diseases but thought that while the Regional Poultry Laboratory workers were studying visceral lymphomatosis, the Cornell workers were studying neural lymphomatosis! This problem continued into the 1960s – consider, for example, the reporting of severe outbreaks of avian leukosis, which it is now recognized was a severe form of MD (Benton *et al.*, 1962; Dunlop *et al.*, 1965), and the transmission studies of what is now known to have been MD but was described by Sevoian and colleagues as lymphotomatosis (Sevoian *et al.*, 1962; Sevoian and Chamberlain, 1963).

The difficulties and confusion reached such a serious state that at the first Conference of the World Veterinary Poultry Association held in Utrecht in 1960, Campbell and Biggs were asked to present papers on the classification of the leukosis complex and fowl paralysis (Biggs, 1961; Campbell, 1961). Both authors considered that the term lymphomatosis was largely responsible for the confusion, and it was agreed by the General Meeting to adopt the proposal of Biggs that the disease originally described by Marek (1907) and called *neurolymphomatosis gallinarum* by Pappenheimer and colleagues (1926, 1929a) be termed 'Marek's disease'. Both authors agreed that leukosis was the term to use for the conditions described by Ellerman (1921), and that the lymphatic form should be termed 'lymphoid leukosis'. As already mentioned, a resolution proposing this classification and terminology was adopted by the first General Meeting of the World Veterinary Poultry Association at its first Conference held in Utrecht in 1960 (Biggs, 1962). This division of the avian leukosis complex into two conditions was largely accepted in Europe, but was greeted with scepticism in the USA for some years.

Evidence for the separation of Marek's disease from lymphoid leukoses

Clinical and pathological studies of field cases

The early evidence that MD was a separate entity from LL came from those who studied the clinical and pathological characteristics of field cases. Although many were not convinced that the diseases could be differentiated, others felt strongly that MD was a different disease from LL and that in most cases a diagnosis could be made of which disease was present in a flock. Diagnosis of individual cases was more difficult and, in some cases, problematical. Some scientists held the view throughout the 1930s and 1940s that MD was a disease entity separate from LL (for example, Gibbs, 1936; Durant and McDougle, 1939, 1945; Fritzsche, 1939). Campbell felt strongly that the Jungherr–Cottrall classification and nomenclature was unsatisfactory because it confused MD with LL, and considered there was adequate clinical and pathological evidence to dissociate these two diseases. He made pleas for the adoption of a classification that would recognize this dissociation, and proposed such a one (Campbell, 1954, 1956). Although his suggested classification was not adopted, his pleas influenced the organizer of the first Conference of the World Veterinary Association to include in the programme both papers and a discussion on this subject (Biggs, 1961; Campbell, 1961). The outcome of these presentations and discussion was the resolution already referred to above, which influenced thinking and the future of MD research.

The evidence for such a dissociation provided by Campbell (1961) and Biggs (1961) differed. In both cases, the support for the separation came from clinical and pathological experience derived from field material. However, Campbell (1961) considered MD to be an inflammatory disease and that only rarely did a neural or visceral lesion become neoplastic, whereas Biggs (1961) considered the early lesions to be inflammatory and the later, more advanced changes to be neoplastic. Biggs (1961) 'concluded that the difference in (1) age group affected, (2) distribution of organs and tissues affected, and (3) histopathogenesis, warrants a distinction in terms for the two diseases'. He also recognized that a problem remained with the diagnosis of visceral lesions, especially if neural lesions were absent.

Transmission studies

Lymphoid leukosis

The absence of conclusive evidence for the concept that there were two diseases was a challenge for the research community. Already there was good evidence that 'visceral lymphomatosis' was transmissible and by a filterable agent, presumably a virus (Burmester, 1947). Later it was shown that the disease could be transmitted using material from field cases of lymphomatosis (Burmester and Fredrickson, 1964). We now know that the disease they were transmitting was LL.

Marek's disease

Many attempts had been made by groups over the years to transmit MD. They suffered from uncertainty of the disease they were working with and either the lack of controls or a significant incidence of the disease in those controls. The most noteworthy of these studies were those of Pappenheimer and colleagues (1926, 1929b), Blakemore (1939), and Durant and McDougle (1945). However, the transmissibility of MD was not generally accepted until Sevoian and colleagues (1962) and Biggs and Payne (1963, 1967) reproduced a high incidence of the disease in inoculated chicks with no (or a very low), incidence in the controls. It was also convincingly shown that the disease could be transmitted by direct and indirect contact between infected and uninfected chicks (Biggs and Payne, 1963). These studies resolved the question of the transmissibility of MD, but not whether this was a different disease from LL.

Experimental studies comparing Marek's disease with lymphoid leukosis

A step forward in answering this question was provided by the studies reported by Biggs and Payne (1964). They compared and contrasted the properties of their strain of MD, HPRS B14, with isolates they had made from natural cases of LL. They studied the disease produced by these agents in a line of chicken susceptible to MD (HPRS RIR) and a line susceptible to LL (L15I WL). The parameters they examined included the incidence of each disease, the median latent period for the development of the relevant disease, the distribution of lesions and tumours, the histology of the lesions, and the ability to infect chick embryo cells and induce a resistance to infection with Rous sarcoma virus (RSV).

The parameters for the two isolates from LL were similar, and also analogous to those described by the Burmester group for RPL 12 virus and other isolates of 'visceral lymphomatosis', but they differed from those for HPRS B14 strain of MD. The incidence of MD was highest in the HPRS RIR line and lowest in the L15I WL line, whereas it was the reverse for LL; the median latent period for LL in both lines of chicken was at least twice that of MD; the distribution of tumours and their frequency in different organs differed; the cytology of the tumour was different; and the leukosis isolates grew in chick embryo cells and induced a resistance to RSV whereas the MD strain did not. It was considered that these studies provided strong evidence for recognizably separate diseases with aetiological specificity for each disease. These results, together with the observation that the MD agent also differs from the leukosis virus in being highly cell-associated (Biggs, 1965, 1966; Biggs and Payne, 1967), led finally to the acceptance of the dissociation of the two diseases in the USA. The report of a committee on Classification and Nomenclature set up during a Technical Workshop of those active in research on the Avian Leukosis Complex in 1965 conceded 'that there should be recognition within the avian leukosis complex of a major division between the two conditions that have been referred to as MD and LL' (Burmester, 1966).

Identifying the Marek's disease virus

The first step towards identifying the virus of MD was the ability to transmit the disease with regularity – i.e. having known infectious material with which to work (Biggs and Payne, 1963, 1967; Sevoian *et al.*, 1962) and the potential for the development of an assay system. Initially the properties of the agent were examined using transmission of overt disease as the criterion of infectivity, but this meant each assay took about 10 weeks and provided a qualitative rather than quantitative result. Because of this limitation, a short-term assay relying on microscopic diagnosis of infection was developed (Biggs and Payne, 1967). This assay took between 2 and 3 weeks, and could be used quantitatively. It enabled more rapid progress, but was time consuming. Using this assay, it was shown in a series of experiments that in both blood and tumour cells infectivity was avidly cell-associated (Biggs and Payne, 1967; Biggs *et al.*, 1968). It was concluded from these studies that the infectious unit was an avian cell. It was assumed that the infectivity could be due to a cell-associated virus; however, it was possible that a cell transplant was responsible for the experimentally produced disease, and this would need negating. Using the sex chromosome as a cell marker it was found that, with rare exceptions, tumour cells were of host and not donor origin in chickens inoculated with blood or tumour cells (Owen and Moore, 1966; Biggs *et al.*, 1968). It was concluded that the infectious agent present in tumour cells and blood was an avidly cell-associated virus, which requires viable cells for the effective transfer of infection. With this knowledge, attempts were made to isolate the virus and develop an assay system in cell culture that would be less laborious than the chick assay.

Identification of Marek's disease herpesvirus

During work directed at the development of a cell culture system for the isolation and assay of the MD agent using tumour cells or whole blood as the inoculum, a cytopathic effect was noted in the UK in cultured chick kidney cells (Churchill and Biggs, 1967; Churchill, 1968) and independently in the USA in cultured duck embryo cells (Nazerian *et al.*, 1968; Solomon *et al.*, 1968). In both cases, a herpes-type virus was seen to be associated with the cytopathic effect.

Circumstantial evidence

Although the herpesvirus was strongly associated with MD in that it was only present in cell cultures showing a characteristic cytopathic effect, and only cultures with this cytopathic effect produced MD when inoculated into chicks (Churchill and Biggs, 1967, 1968; Churchill, 1968; Nazerian *et al.*, 1968; Solomon *et al.*, 1968), it was possible that it could be a passenger virus commonly found in chicken tissue. It is difficult to fulfil Koch's postulates for a virus, because it is almost impossible to guarantee its purity in cell culture and in cell-free preparations. For a cell-associated virus, the postulates are impossible to fulfil. The burden of proof of the identity of the isolated herpesvirus and the infectious agent of MD therefore could only come from accumulated circumstantial evidence.

In addition to the studies detailed above, the herpesvirus was isolated from all laboratory strains of the infectious agent examined and consistently from tissues from chickens with MD from a number of farms, both in the UK and the USA. Using chicks for assaying the MD agent and cytopathic effect, and plaque production in cell culture for assaying the herpesvirus, it was found that: 1. infectivity of both tumour and blood cells and herpesvirus-infected cells in culture are cell-associated, whether assayed in chicks or cell culture; 2. both infected blood and tumour cells and infected cell culture cells respond quantitatively in the same way to treatments destroying infectivity, whether assayed in chicks or cell culture; 3. there was a correlation between the ability to produce the characteristic cytopathic effect in cell culture and MD in chicks (Churchill and Biggs, 1967, 1968; Biggs *et al.*, 1968; Witter *et al.*, 1969). Although all working with MD accepted this evidence for the identity of the herpesvirus as the causative agent of MD, there were still some scientists who were sceptical.

Definitive evidence

Two more developments were required to provide proof of the hypothesis that the herpesvirus was the causative agent of MD and to satisfy the doubters. Following the demonstration of large quantities of viral antigen in the feather-follicle epithelium of infected chickens associated with the fully-enveloped virus, studies showed that cell-free virus produced by the disruption of feather-follicle epithelial cells could reproduce the disease (Calnek and Hitchner, 1969; Calnek *et al.*, 1970a). These findings also provided an explanation for the paradoxical contagious nature of the disease. The second development concerned vaccines against the disease. Passage of infected chick kidney cell cultures resulted in attenuation of the pathogenicity of such cells for chickens, and at the same time altered the antigenic attributes of the herpesvirus and the plaque type produced. Cells infected by the attenuated virus could be used to immunize chickens against the disease, both under experimental and field conditions (Churchill *et al.*, 1969a, 1969b; Biggs *et al.*, 1970). Soon after, a virus related to the MDV, which had been isolated from a turkey (herpesvirus of turkeys), was shown to immunize chickens against the disease (Okazaki *et al.*, 1970; Purchase *et al.*, 1972). The lyophilization of this virus (Calnek *et al.*, 1970b) in high enough titre for use as a vaccine resulted in cell-free herpesvirus of turkeys being available on the market as an effective vaccine. The demonstration that an attenuated MD herpesvirus and the antigenically related herpesvirus of turkeys were able to immunize chickens against MD provided conclusive evidence that the herpesvirus was the causative agent of the disease.

Summary

Subsequent developments have fully supported the view that MD is a disease entity separate from LL, and that the herpesvirus is the causative virus of the disease. This virus is now generally referred to as the *Marek's disease virus* (MDV).

It is remarkable that it took 60 years from the first description of MD for it to be recognized universally as a distinct entity and for its causative agent to be isolated and understood, especially as the concept that it was an infectious disease probably due to a virus had been present since the 1920s. However, once successful, repeatable and convincing transmission of the disease had been achieved progress was impressively rapid – albeit in difficult circumstances – in the establishment of the aetiology and prevention of the disease.

References

Benton, W. J., Cover, M. S. and Krauss, W. C. (1962). *Avian Dis.*, **6**, 430–435.

Biggs, P. M. (1961). *Br. Vet. J.*, **117**, 326–334.

Biggs, P. M. (1962). In *Proceedings of the Thirteenth Symposium of the Colston Research Society, Vol. XIII of the Colston Papers*, pp. 83–99. Butterworth, London.

Biggs, P. M. (1965). In *Proceedings of the 3rd Congress of the World Veterinary Poultry Association, Paris*, pp. 61–67, WVPA, Paris.

Biggs, P. M. (1966). In *Proceedings of the 13th World's Poultry Congress, Kiev*, pp. 91–118, WPSA, Kiev.

Biggs, P. M. and Payne, L. N. (1963). *Vet. Rec.*, **75**, 177–179.

Biggs, P. M. and Payne, L. N. (1964). *Natl Cancer Inst. Monograph*, **17**, 83–97.

Biggs, P. M. and Payne, L. N. (1967). *J. Natl Cancer Inst.*, **39**, 267–280.

Biggs, P. M., Churchill, A. E., Rootes, D. G. and Chubb, R. C. (1968). *Persp. Virol.*, **6**, 211–230.

Biggs, P. M., Payne, L. N., Milne, B. S. *et al.* (1970). *Vet. Rec.*, **87**, 704–709.

Blakemore, F. (1939). *J. Comp. Path.*, **52**, 144–159.

Burmester, B. R. (1947). *Cancer Res.*, **7**, 786–797.

Burmester, B. R. (1966). *Poultry Sci.*, **45**, 1411–1415.

Burmester, B. R. and Fredrickson, T. N. (1964). *J. Natl Cancer Inst.*, **32**, 37–63.

Calnek, B. W. and Hitchner, S. B. (1969). *J. Natl Cancer Inst.*, **42**, 935–949.

Calnek, B. W., Adldinger, H. K. and Kahn, D. E. (1970a). *Avian Dis.*, **14**, 219–233.

Calnek, B. W., Hitchner, S. B. and Adldinger, H. K. (1970b). *Appl. Microbiol.*, **20**, 723–726.

Campbell, J. G. (1945). *J. Comp. Path.*, **55**, 308–321.

Campbell, J. G. (1954). In *Proceedings of the Xth World's Poultry Congress, Edinburgh*, pp. 193–197, Department of Agriculture for Scotland, Edinburgh.

Campbell, J. G. (1956). *Vet. Rec.*, **68**, 527–529.

Campbell, J. G. (1961). *Br. Vet. J.*, **117**, 316–325.

Chubb, L.G. and Gordon, R. F. (1957). *Vet. Rev. Annot.*, **3**, 97–120.

Churchill, A. E. (1968). *J. Natl Cancer Inst.*, **41**, 939–950.

Churchill, A. E. and Biggs, P. M. (1967). *Nature, Lond.*, **215**, 528–530.

Churchill, A. E. and Biggs, P. M. (1968). *J. Natl Cancer. Inst.*, **41,** 951–956.

Churchill, A. E., Chubb, R. C. and Baxendale, W. (1969a). *J. Gen. Virol.*, **4,** 557–564.

Churchill, A. E., Payne, L. N. and Chubb, R. C. (1969b). *Nature, Lond.*, **221,** 744–747.

Cole, R. K. and Hutt, F. B. (1951). *Poultry Sci.*, **30,** 205–212.

Cottrall, G. E. (1952). In *Proceedings of the 89th Meeting of the American Veterinary Medicine Association*, Atlantic City, pp. 285–293, AVMA, Chicago.

Cottrall, G. E., Burmester, B. R. and Waters, N. F. (1954). *Poultry Sci.*, **33,** 1174–1184.

Dunlop, W. R., Kottaridis, S. D., Gallagher, J. R. *et al.* (1965). *Poultry Sci.*, **44,** 1537–1540.

Durant, A. J. and Mc Dougle, H. C. (1939). *Missouri Agric. Exp. Sta. Bull.*, **304,** 3–23.

Durant, A. J. and Mc Dougle, H. C. (1945). *Missouri Agric. Exp. Sta. Bull.*, **393,** 1–18.

Ellerman, V. (1921). *The Leucosis of Fowls and Leucaemia Problems*. Gyldendal, London.

Fritzsche, K. (1939). *Z. Infektkrankh. Haustiere*, **55,** 68–74.

Gibbs, C. S. (1936). *Massachusetts Agric. Exp. Sta. Bull.*, **337,** 1–31.

Hutt, F. B. (1951). *Wlds Poult Sci. J.*, **7,** 16–25.

Jungherr, E. (1941). *Am. J. Vet. Res.*, **3,** 116.

Kaup, B. F. (1921). *Am. Ass. Instr. Invest. Poult. Husb.*, **7,** 15.

Marek, J. (1907). *Dtsch Tierärztl. Wochenschr.*, **15,** 417–421.

Nazerian, K., Soloman, J. J., Witter, R. L. and Burmester, B. R. (1968). *Proc. Soc. Exp. Biol. Med.*, **127,** 177–182.

Owen, J. J. T. and Moore, M. A. S. (1966). *J. Natl Cancer Inst.*, **37,** 199–209.

Okasaki, W., Purchase, H. G. and Burmester, B. R. (1970). *Avian Dis.*, **14,** 413–429.

Pappenheimer, A. M., Dunn, L. C. and Cone, V. (1926). *Storrs Agric. Exp. Sta. Bull.*, **143,** 186–290.

Pappenheimer, A. M., Dunn, L. C. and Cone, V. (1929a). *J. Exp. Med.*, **49,** 63–86.

Pappenheimer, A. M., Dunn, L. C. and Siedlin, S. M. (1929b). *J. Exp. Med.*, **49,** 87–102.

Purchase, H. G., Okazaki, W. and Burmester, B. R. (1972). *Avian Dis.*, **16,** 57–71.

Sevoian, M. and Chamberlain, D. M. (1963). *Avian Dis.*, **7,** 97–102.

Sevoian, M., Chamberlain, D. M. and Counter, F. (1962). *Vet. Med.*, **57,** 500–501.

Solomon, J. J., Witter, R. L., Nazerian, K. and Burmester, B. R. (1968). *Proc. Soc. Exp. Biol. Med.*, **127,** 173–177.

van der Walle, N. and Winkler-Junius, E. (1924). *Tijdschr. Vergelijk. Geneesk. Gezondhleer*, **10,** 34–50.

Waters, N. F. (1954). *Poultry Sci.*, **33,** 365–373.

Witter, R. L., Burgoyne, G. H. and Solomon, J. J. (1969). *Avian Dis.*, **13,** 171–184.

The genome content of Marek's disease-like viruses

3

KLAUS OSTERRIEDER* and **JEAN-FRANÇOIS VAUTHEROT****

*Department of Microbiology and Immunology, College of Veterinary Medicine, Cornell University, Ithaca, New York, USA
**Centre de Recherches de Tours, INRA, Nouzilly, France

Introduction

Three members of the genus *Mardivirus*, Marek's disease herpesvirus serotypes 1 and 2 (MDV-1 and MDV-2) as well as serotype 3 or herpesvirus of turkeys (HVT) are recognized (Davison *et al.*, 2002a). The *Mardivirus* genus was established because accumulated genomic information clearly indicated their affiliation with the *Alphaherpesvirinae* subfamily of the *Herpesviridae*, but distinguished them from the other genera of the subfamily, the *Simplex-*, *Varicello-* and *Laryngoviruses*. More than 120 viruses have been assigned to the *Herpesviridae*, and most vertebrate animals harbour more than one member of this large virus family (Davison, 2002a, 2002b; Davison *et al.*, 2002a). The genomes of all *Herpesviridae* are double-stranded linear DNA molecules that range in size from 108 to 230 kbp (Maotani *et al.*, 1986; Davison, 2002b). A total of six different general genome organizations, referred to as classes A through F, are distinguished in the *Herpesviridae* (Roizman, 1996). Only class D and E genomes are found in the *Alphaherpesvirinae*, and MDV-1, MDV-2 and HVT represent class E genomes, an organization that is identical to that of the prototype representative of the virus subfamily, herpes simplex virus type 1 (HSV-1) (Roizman, 1982, 1996). Class E genomes comprise two unique sequences, a long (unique-long, U_L) and a short (unique-short, U_S), each of which is bracketed by inverted internal (IR_L, IR_S) and terminal repeats (TR_L, TR_S: Fig. 3.1). The TR_L and IR_L as well as the TR_S and IR_S are identical in sequence and present in inverse orientations. During DNA replication, the U_L and U_S regions can flip-flop relative to the other unique region, and consequently four isomeric forms of the viral DNA exist, probably in equimolar amounts (Roizman, 1982; Roizman and Sears, 1996).

Recently, the entire genomes of representatives of all three members of the *Mardivirus* genus have been sequenced. These studies have shown that the gene contents and linear arrangements of the three viruses are similar in general, but vary considerably with regard to guanine and cytosine (GC) content and size. Whereas the GC content of MDV-1 is as low as 44.1 per cent, that of MDV-2 is 53.6 per cent. Interestingly, HVT has a GC content that lies in between these two extremes (47.2 per cent) (Lee *et al.*, 2000; Tulman *et al.*, 2000; Afonso *et al.*, 2001; Izumiya *et al.*, 2001; Kingham *et al.*, 2001). A total of 99 (HVT), 102 (MDV-2) and 103 (MDV-1) genes have been clearly identified, and the vast majority of the gene repertoire of all three viruses consists of open reading frames (ORF) that are homologous to genes found in other *Alphaherpesvirinae*. Not unexpectedly, genus- and type-specific genes are also present in the three genomes. Among these, the most prominent are the MDV-1-specific *meq* (MDV EcoRI-Q) and the *pp38* genes that both have been implicated in MDV latency and tumour formation (Xie *et al.*, 1996; Ross *et al.*, 1997). In addition, a chemokine-encoding gene *vIL8* (viral interleukin-8) is expressed by MDV-1, but not the avirulent MDV-2 or HVT genomes (Parcells *et al.*, 2001). On the other hand, the most prominent HVT-specific ORF, *NR-13*, encodes a putative *Bcl*-2 homologue that exhibits anti-apoptotic properties, but no homologue has been identified in either MDV-1 or MDV-2 (Kingham *et al.*, 2001; Aouacheria *et al.*, 2003).

Besides providing important information about the exact base composition and gene content of the individual viruses, the sequence information on the three known *Mardiviruses* allows two important conclusions. First, the three viruses certainly represent individual and clearly distinct virus species that are only misleadingly referred to as 'serotypes'. The rather large variation in base composition and genome size, as well as the relatively low homology of individual proteins that reaches a maximum identity of 88 per cent between any of the three viruses, clearly suggests an independent but probably parallel evolution of separate and different virus species. Secondly, it is becoming increasingly clear that tumour formation, which only MDV-1 is capable of causing, is probably dependent on a complex interplay of a number of gene products and regulatory elements or even RNA structures (Fragnet *et al.*, 2003). We propose here that the nomenclature of the members of the *Mardivirus* genus be changed, with only MDV serotype 1 referred to as MDV, since this virus actually causes Marek's disease. We will use the official nomenclature of '*Gallid Herpesvirus* Type 3 (GaHV-3)' for the closely related MDV-2 from here on.

Gene content and transcription of the *Mardiviruses*

As stated above, the majority of the genes encoded by MDV, GaHV-3, and HVT are homologous to those encoded by other members of the *Alphaherpesvirinae*.

Largely by analogy to and deduction from the better studied members of the subfamily, especially HSV-1, 'putative' functions have been attributed to genes and gene products that exhibit reasonably high homology to the HSV-1 genes (Kingham *et al.*, 2001). It has to be stated, first, that gene function can vary, even if the encoded proteins exhibit high sequence or structural homology, and that any prediction of gene function based on sequence information is limited. Secondly, the temporal expression of certain genes within the *Alphaherpesvirinae* is variable to a certain degree. To give one important example, HSV-1 expresses a total of six immediate-early (IE, α) genes, whereas a sole IE protein is produced after infection of susceptible cells with equine herpesvirus type 1 (EHV-1) that is capable of inducing the cascade-like expression of viral genes throughout the replicative cycle (Gray *et al.*, 1987; Robertson *et al.*, 1988).

The initiation and the temporal regulation of MDV gene expression are less well known. Owing to the highly cell-associated nature of MDV, high multiplicity and simultaneous infection of cultured cells is difficult; therefore the temporal regulation of MDV gene expression is difficult to analyse, and mainly latently-infected or tumour cells are used, in which the lytic cycle can be induced by various chemicals such as iodo-deoxyuridine (IUDR), 12-O-tetradecanoyl phorbol-13-acetate (TPA) or N-butyrate (Calnek *et al.*, 1984). In early attempts to establish a map of MDV gene expression, a total of seven to eight IE, two early (E, β) and twelve late (L, γ) genes were identified and the kinetics of mRNA expression appeared to be dependent on the virus strain and cell type used (Maray *et al.*, 1988). In a later study, 66 MDV-specific transcripts were characterized, eleven of which appeared to be expressed with IE kinetics as determined using cycloheximide to prevent *de novo* protein synthesis, and six with E kinetics as demonstrated using the inhibitor of MDV DNA replication FMAU (Schat *et al.*, 1989). In the case of GaHV-3, a transcriptional map covering approximately 30 per cent of the genome is available; however, information on the kinetic classes of individual transcripts is very limited (Izumiya *et al.*, 1998, 1999; Jang *et al.*, 1998; Tsushima *et al.*, 1999). Because the genomic sequences available for the *Mardiviruses* suggest production of approximately 100 proteins in each of the three viruses, it is clear that there is a myriad of unidentified transcripts in cells infected with MDV, GaHV-3 or HVT. In addition, it is clear that the repertoire of transcripts differs enormously between lytically and latently infected or transformed cells (Cantello *et al.*, 1994, 1997; Qian *et al.*, 1995; Ross *et al.*, 1993, 1997; Ross, 1999), and that transcription is cell-type specific to a certain degree. Although MDV has been shown to infect at least four different cell types *in vivo*, including B cells, T cells, epithelial cells of the feather follicle and the kidneys, and macrophages (Calnek, 2001; Barrow *et al.*, 2003), most information on MDV transcription has been gathered from cultured chicken embryo cells (CEC), chicken kidney cells or transformed T cells, in which the lytic cycle was chemically induced (Cantello *et al.*, 1997; Parcells *et al.*, 1999, 2003). In the following paragraphs, we shall try to summarize the current knowledge on the kinetics of MDV gene expression.

In members of the *Alphaherpesvirinae* that are closely related to the *Mardiviruses*, the initiation of lytic gene expression starts with the binding of the U_L48 homologous protein, also named α-transinducing factor (α-TIF) or VP16, to a TAATGARAT motif located upstream of the ICP4 ORF (Figure 3.1). This binding leads to the recruitment of cellular transcription factors and finally to the expression of the ICP4 and other IE proteins (Roizman and Sears, 1996). The presence of α-TIF, however, is not absolutely required to start transcription, because herpesviral DNA has been shown to be infectious *per se* (Roizman, 1996). In addition, the U_L48 homologues of both varicella zoster virus (VZV) and MDV are dispensable for virus growth in cultured cells (Cohen and Seidel, 1994; Dorange *et al.*, 2002), although nothing is known about the requirements for VZV or MDV α-TIF homologues to allow virus replication *in vivo*. Out of the candidate IE genes, the kinetic class of only the ICP27 homologue of MDV has been identified, and it was shown that ICP27 is produced from a true IE transcript. In addition to this 'classical' IE gene, the gK gene (U_L53) that is the direct neighbour of the ICP27 gene (U_L54) appears to be expressed with IE kinetics (Ren *et al.*, 1994), which is surprising since gK is expressed as a true-late (γ2) protein in all *Herpesviridae* analysed so far. Another gene expressed with IE kinetics is the MDV-1 specific L-ORF10 gene (MDV006/MDV075) (Hong and Coussens, 1994; Lee *et al.*, 2000; Tulman *et al.*, 2000). The highly phosphor-ylated protein is expressed from two different splice variants that share identical carboxy-terminal ends, and was detected in cells lytically infected with oncogenic and attenuated MDV. It localizes mainly to the cytoplasmic fraction of infected cells, and was reported to be also expressed in some latently infected and trans-formed cells (Hong and Coussens, 1994; Hong *et al.*, 1995; Ui *et al.*, 1998); however, the function of this unique protein has not been determined so far.

With regard to IE gene expression, much research has concentrated on the MDV ICP4 homologue, because herpesviral ICP4 homologues are potent IE transactivators of early gene expression, and are able – in a time-dependent manner – to modulate their own expression by recruiting various cellular and viral transcription factors as infection progresses (Smith *et al.*, 1992, 1995; Roizman and Sears, 1996; Roizman, 1999). However, it has not been formally proven that the MDV ICP4 homologue, which is located in the IR_S and TR_S region of the genome, encodes a protein that is expressed with IE kinetics. First described as a 1415 amino acid protein (Anderson *et al.*, 1992), it became obvious later on that MDV (as well as GaHV-3 and HVT) encodes a much larger version of the putative transactivator of early genes with a theoretical coding capacity for an ICP4 protein of 2321 amino acids in size (Tulman *et al.*, 2000). From the ICP4 genomic region at least five sense transcripts, two of which are spliced, have been identified, ranging in size from 6.2 to 10 kb, and there is a wealth of antisense transcripts to MDV ICP4 that are detectable in latently infected and/or tumour cells (Cantello *et al.*, 1994, 1997; Li *et al.*, 1994; McKie *et al.*, 1995; Xie *et al.*, 1996; Ross *et al.*, 1997; Ross, 1999; Morgan *et al.*, 2001). Using poly- and monoclonal ICP4-specific antibodies, four (possibly) ICP4

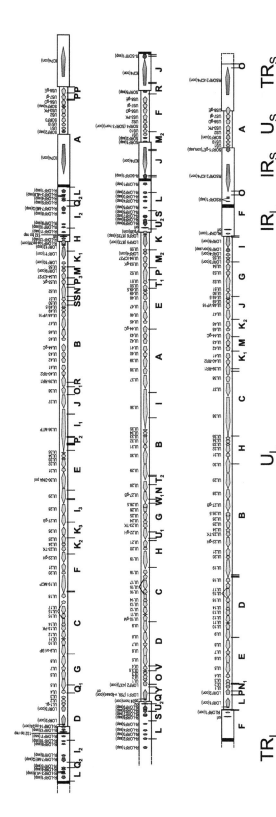

Figure 3.1 Genomic organization of the MDV, GaHV-3, and HVT modified according to Kingham *et al.* (2002). Shown are MDV in the upper bar, GaHV-3 (MDV-2) in the middle bar, and HVT (lower bar). The terminal and internal repeat long regions (TR_L, IR_L), the unique long region (U_L) as well as the internal and terminal repeat short regions (IR_S, TR_S) and the unique short region (U_S) are shown. Open reading frames (ORFs) given in light grey have homologues in HSV-1. ORFs in dark grey and labelled (con) are conserved between the three *Mardiviruses*, and ORFs in dark grey and labelled (ssp) are strain-specific and only found in the respective virus.

protein moieties of 210, 155, 140 and 80 kDa transcribed from this large ORF have been identified (Xing *et al.*, 1999), yet information on the function of the different polypeptides, especially with regard to transactivation of early promoters, self-regulation or their temporal appearance, is fragmented. It was shown that ICP4 may activate the early gene products pp24 and pp38, but alone it is not sufficient to initiate a complete lytic cycle from latently-infected or transformed cells (Pratt *et al.*, 1994).

Despite the report of as many as six E MDV genes, the only E MDV genes identified so far are MDV008 (R-LORF-14) and MDV073 (R-LORF14a) expressing the closely related phosphoproteins pp24 and pp38 (Zhu *et al.*, 1994). The two proteins share the amino-terminal 65 amino acids that are encoded by repeated sequences, but differ in their carboxy-terminal portions (Lee *et al.*, 2000; Tulman *et al.*, 2000). Homologues of pp24 and pp38 are found in GaHV-3 and HVT (Afonso *et al.*, 2001; Izumiya *et al.*, 2001; Kingham *et al.*, 2001), and the expression of the two phosphoproteins, which are believed to form a complex in infected cells, is one of the hallmarks of the lytic infection cycle (Baigent *et al.*, 1998; Parcells *et al.*, 1999, 2003). Deletion of pp38 from the very virulent MDV strain Md5 resulted in a virus that was severely impaired in early lytic replication in the chicken, and consequently was virtually unable to cause visceral tumours or nerve lesions, while replication *in vitro* was unaffected (Reddy *et al.*, 2002). Whether pp24, pp38 or both related proteins play any role in the regulation of MDV L gene expression is questionable, since both proteins are found predominantly in the cytoplasm of infected cells. Both pp24 and pp38 are transcribed from a bi-directional promoter that also harbours a lytic origin of replication (ori_{Lyt}), from which major latency-associated transcripts, the 1.8-kb family of transcripts, are initiated (Shigekane *et al.*, 1999). However, the transcriptional organization and the gene content in this region, present in both long repeats of the MDV genomes, are very confusing. Whereas Md5 appears to encode a 115 amino acid protein (MDV007) that is expressed from an mRNA directed antisense to pp24/pp38 (Tulman *et al.*, 2000), Parcells *et al.* (2003) recently reported on the expression of a putative highly basic 124 amino acid protein called Hep that partially overlaps with pp24/pp38 and would be expressed from three different RNAs, one of which also contains the pp24/pp38 message. So far, nothing is known about the temporal regulation of the expression of MDV enzymes and nucleic acid-binding proteins involved in *de novo* DNA synthesis. These include the genes such as thymidine kinase (MDV036; U_L23), dUTPase (MDV063; U_L50), ribosyl reductase subunits (MDV052 and MDV053; U_L39 and U_L40), origin-binding protein (MDV021, U_L9), DNA polymerase subunits (MDV043 and MDV055; U_L30 and U_L42), proteins forming the helicase-primase complex (MDV017, MDV020, and MDV066; U_L5, U_L8, and U_L52), and single-stranded DNA binding protein (MDV042; U_L29). It is reasonable to assume that they are regulated with early kinetics, because these proteins are very likely to be essential for viral DNA synthesis and the transcription of late structural genes (Tulman *et al.*, 2000).

It is generally accepted that – similar to the situation with other *Herpesviridae* – the majority of structural proteins of MDV are products of L genes, although at least some IE and E proteins (i.e. transactivating factors like the HSV-1 proteins, ICP0 and ICP4) have been shown to be structural components of the mature virion (Roizman, 1996; Roizman and Sears, 1996). In the case of *Mardiviruses*, the identification of structural proteins is not trivial because the virus is highly cell-associated *in vivo* and *in vitro*. Therefore, purification of extracellular virions is difficult if not impossible. However, it is known that MDV membrane (glyco)proteins, with the exception of gK, are expressed from L genes. As stated above, the gK transcript – in contrast to the gK transcript in all other *Herpesviridae* studied so far – was reported to be regulated as an IE gene, because it is detectable in the presence of cycloheximide (Ren *et al.*, 1994). In addition, the major capsid protein VP5 and the major tegument proteins expressed from the MDV genes MDV059–MDV062 that are homologous to the HSV-1 U_L46 to U_L49 genes encoding VP11/12, VP13/14, VP16 (α-TIF) and VP22, have been shown to be expressed with late kinetics (Dorange *et al.*, 2000; Lupiani *et al.*, 2001). A later study has also demonstrated that VP11/12, VP13/14 and VP16 are dispensable for MDV growth in cultured cells, whereas expression of VP22 is absolutely required for virus replication, and infection cannot progress from an infected cell to neighbouring uninfected cells in the absence of VP22 (Dorange *et al.*, 2002) (Table 3.1). This is surprising, since VP22 is dispensable for the growth of closely related viruses like HSV-1, pseudorabies virus or EHV-1 (del Rio *et al.*, 2002). The requirements for major tegument proteins in MDV replication – at least with regard to VP16 and VP22 – appear to be very different from those identified for other *Herpesviridae*, with the notable exception of VZV, in which the VP16 homologue, the gene 10 product, was also shown to be non-essential for virus growth (Cohen and Seidel, 1994).

The similarities between highly cell-associated *Alphaherpesvirinae* VZV and MDV in the requirement for certain gene products that are involved in cell-to-cell spread – the predominant means of virus replication *in vitro* and *in vivo* – are also evident when the functions of late glycoprotein genes are investigated. Neither VZV nor MDV encode for a gG or gJ homologue, and the gD-encoding gene (U_S6) of MDV, which is absent from VZV, is non-essential for growth, efficient horizontal spread and tumour formation (Parcells *et al.*, 1994; Anderson *et al.*, 1998) (Table 3.1). In contrast, gE and gI, which form a non-covalently linked complex in infected cells, are absolutely required for MDV replication (Schumacher *et al.*, 2001). These findings are consistent with those reported for VZV, in which the deletion of gE or gI results in viruses that are massively impaired in their growth in cultured cells (Cohen and Nguyen, 1997; Mallory *et al.*, 1997, 1998) (Table 3.1). Future research will need to concentrate on the analysis of the structural requirements for efficient MDV replication, i.e. on the proteins involved in virus assembly, maturation, egress, and direct cell-to-cell spread.

Table 3.1 MDV mutants generated by RecE/T cloning.

Mutant designation	Gene/protein[1]	Growth in cultured CEC[2]	Growth of the corresponding HSV-1 mutant[3]	Reference
BAC20[4]		+++	n.a.	Schumacher et al., 2000
ΔMDV010	Viral lipase	++	n.a.	Kamil et al. (unpublished)
ΔU$_L$1	gL	−	−	Schumacher et al. (unpublished)
ΔU$_L$10	gM	−	++	Tischer et al., 2002b
ΔU$_L$11	Myristylated tegument protein	−	++	Schumacher et al. (unpublished)
ΔU$_L$20	Membrane protein	−	++	Schumacher et al. (unpublished)
ΔU$_L$22	gH	−	−	Schumacher et al. (unpublished)
ΔU$_L$27	gB	−	−	Schumacher et al., 2000
ΔU$_L$31	Nuclear phosphoprotein	−	−	Tischer et al. (unpublished)
ΔU$_L$34	Membrane protein	−	−	Tischer et al. (unpublished)
ΔU$_L$35	VP26	+	++	Tischer et al. (unpublished)
ΔU$_L$37	Tegument protein	−	−	Schumacher et al. (unpublished)
ΔU$_L$41	VHS	+++	+++	Schumacher et al. (unpublished)
ΔU$_L$43	Membrane protein	−	?	Schumacher et al. (unpublished)
ΔU$_L$44	gC	++++	++	Tischer et al. (unpublished)
ΔU$_L$45	Membrane protein	+++	+++	Schumacher et al. (unpublished)
ΔU$_L$46	Tegument, VP11/12	++	++	Dorange et al., 2002
ΔU$_L$47	Tegument, VP13/14	++	++	Dorange et al., 2002
ΔU$_L$48	Tegument, VP16	++	−	Dorange et al., 2002
ΔU$_L$46–48	Tegument	+	n.a.	Dorange et al., 2002
ΔU$_L$49	Tegument, VP22	−	+	Dorange et al., 2002
ΔU$_L$48–49	VP16, VP22	−	n.a.	Dorange et al., 2002

Table 3.1 (*contd.*)

Mutant designation	Gene/protein[1]	Growth in cultured CEC[2]	Growth of the corresponding HSV-1 mutant[3]	Reference
$\Delta U_L 49.5$	Membrane protein	–	++	Tischer *et al.*, 2002b
$\Delta U_L 51$	Putative virion phosphoprotein	+	+	Schumacher *et al.* (unpublished)
$\Delta U_L 53$	gK	–	+	Schumacher *et al.* (unpublished)
ΔMDV071	VZV ORF2	+++	n.a.	Schumacher *et al.* (unpublished)
$\Delta U_S 3$	Protein kinase	+/(+++)[5]	++	Schumacher *et al.* (unpublished); Sakaguchi *et al.*, 1993
$\Delta U_S 7$	gI	–	+++	Schumacher *et al.*, 2001
$\Delta U_S 8$	gE	–	+++	Schumacher *et al.*, 2001
$\Delta U_S 7$–$U_S 8$	gI–gE	–	++	Schumacher *et al.*, 2001

[1]Genes and gene products are named in accordance to those in HSV-1.
[2]Mutants were analysed in cultured chicken embryo cells (CEC). Virus growth is given semiquantitatively from no growth at all (–) to unaffected growth (+++).
[3]As summarized by Roizman (1996).
[4]All mutants are based on the BAC clone of strain 584Ap80C (BAC20) (Schumacher *et al.*, 2001).
[5]Sakaguchi *et al.* (1993) reported unaffected growth of the mutant; however, Schumacher *et al.* (unpublished) found a 50 per cent reduction in plaque size.

Transcription and gene expression during MDV latency

Latency in the *Alphaherpesvirinae* appears to be regulated and controlled predominantly by transcripts and proteins initiating in, and expressed from, the repeat regions of the genomes. Although the main players in the process have been identified, the mechanisms of the establishment and the maintenance of latency are still one of the enigmas of herpesvirus research (Preston, 2000). In the case of MDV, it was proposed that the IE protein ICP4 and the E proteins pp38 and/or pp24 are involved in the maintenance of MDV latency and/or transformation because oligonucleotides antisense to ICP4 and pp38/24 inhibited proliferation of MDV tumour cells and soft agar colony formation (Xie *et al.*, 1996). In addition, the transcripts antisense to ICP4 have also been implicated – largely by analogy to what is known for other *Herpesviridae* – in the establishment and maintenance of MDV latency. A number of so-called latency associated transcripts (LAT) that are antisense to the ICP4 ORF are detectable in both lytically infected and, mainly, MDV-derived tumour cells (Cantello *et al.*, 1994, 1997; Li *et al.*, 1994). Morgan and co-workers generated a mutant virus derived from the very virulent RB1B strain that had a lacZ gene inserted into the site where all LAT antisense to ICP4 are initiated (Cantello *et al.*, 1997; Morgan *et al.*, 2001). The resulting LAT virus mutant was unable to cause any tumours in inoculated chickens, although lytic virus replication *in vitro* appeared unaffected (Morgan *et al.*, 2001). Also located in the repeat regions of the MDV genome (TR$_L$ and IR$_L$) is the meq gene that encodes one of the most abundantly expressed proteins during latency; this is more fully described in Chapter 4.

Another region that is clearly transcriptionally active during latent infection or in transformed T cells is located in close proximity to the border between the U$_L$ and TR$_L$/IR$_L$ regions of the genomes. From this region the so-called 1.8-kb family of transcripts is initiated, and the transcripts and/or the putative proteins encoded by these transcripts have been shown to be involved in the induction and maintenance of MDV latency and transformation (Kawamura *et al.*, 1991; Hong and Coussens, 1994; Peng *et al.*, 1994; Hong *et al.*, 1995; Hayashi *et al.*, 1999). The genomic region from which the 1.8-kb family of transcripts originates is highly complex, and the literature on the subject reflects this. It is clear, however, that the transcripts are controlled by a bi-directional promoter, which overlaps the ori$_{Lyt}$ of MDV (Kopacek *et al.*, 1993; Shigekane *et al.*, 1999). Some of the transcripts contain the 132-bp repeats, a genomic peculiarity of MDV, which were shown to increase in numbers with serial propagation of MDV strains in cultured cells. As a consequence they were considered to be involved in virulence, although recent evidence suggests that the 132-bp repeats are not directly responsible for virus replication and/or tumour formation capabilities (Maotani *et al.*, 1986; Silva *et al.*, 2003). Coincidentally with the amplification of the direct repeats, the size of the 1.8-kb transcripts is reduced (Bradley *et al.*, 1989); however, the consequences of the modification of the transcripts originating in this region on latency and/or tumour formation have not been addressed by the generation and testing of respective mutant viruses. It has been shown that at least two proteins are expressed from the 1.8-kb family of transcripts: a 7-kDa protein and the 14-kDa IE protein (Hong and Coussens, 1994; Peng *et al.*, 1994). Their roles in the MDV life cycle or their importance for latency and tumour formation have yet to be determined, but at least the 14-kDa protein has been reported to be expressed in the cytoplasm of cells that were lytically – or latently – infected with MDV. In addition, the pp14 phosphoprotein was also detectable in the MSB-1 tumour cell line (Hong and Coussens, 1994; Hong *et al.*, 1995). The expression of the protein in all the different stages of virus infection, as well as its localization in the cytoplasm, have precluded speculation on its function and functional importance for the establishment and maintenance of (or reactivation from) latency.

Manipulation of the MDV genome: new strategy using bacterial artificial chromosomes

In 1997, Messerle and co-workers applied bacterial artificial chromosome (BAC) cloning and mutagenesis to the more than 220-kb murine cytomegalovirus genome (Messerle *et al.*, 1997). These authors succeeded in establishing an entire infectious herpesviral genome as a single copy mini-F plasmid in *Escherichia coli*. In 2000, BAC technology was introduced to MDV, and successful virus reconstitution in cultured cells of an avirulent virus from cloned DNA was reported by Schumacher *et al.* (2000). Later, the widely-used vaccine strain CVI988 (Rispens) and the very virulent RB1B strain were cloned as infectious BAC (Petherbridge *et al.*, 2003). The establishment of herpesviral genomes, including those of MDV, as BAC has several enormous advantages. First, the viral genomes are fixed at a certain stage in *E. coli* and can be maintained independently of propagation in eukaryotic cells. Secondly, the low copy number (usually one to two copies) of mini-F plasmids favours genetic stability. Thirdly, MDV BAC clones are amenable for an application of the powerful recombination machinery of *E. coli* (Brune *et al.*, 2000).

Principally, two methods of mutagenesis are used to delete or modify MDV genes in BAC clones: RecA and RecE/T mutagenesis or cloning. RecE/T cloning is a relatively novel mutagenesis method, and it has been used as a powerful tool to remove or disrupt any desired gene from various BAC clones. This method is a one step mutagenesis with linear DNA fragments containing a selectable marker and short homologous sequences flanking the target sequence, and it has been adopted and optimized for the mutagenesis of MDV BACs (Schumacher *et al.*, 2001; Tischer *et al.*, 2002a, 2002b) (Table 3.1). RecE/T cloning requires the recombinogenic functions that are encoded by recE and recT from the prophage Rac or by their functional homologues redα and redβ from the bacteriophage λ. RecE is a $5'$-$3'$ exonuclease and RecT is a single-stranded DNA (ssDNA)-binding protein that promotes ssDNA annealing, strand transfer, and strand invasion *in vitro* (Muyrers *et al.*, 1999, 2000, 2001; Zhang *et al.*, 2000). To prevent degradation of the linear DNA in bacteria, either exonuclease-negative bacteria are used or the RecBCD exonuclease inhibitor Redγ (*gam*) from bacteriophage λ is expressed. The RecE/T or redα/β recombination allows the introduction of selectable markers, usually an antibiotic resistance gene, with homologous sequences as short as 25–50 nucleotides by a double crossing-over event. In contrast to shuttle or RecA mutagenesis, which allows the introduction of any kind of mutation (e.g. deletion, point mutation, insertion or sequence replacement) without leaving any unintended traces, RecE/T cloning requires the insertion of a selectable marker. Even though the selectable marker can be removed by using flanking Flp target recognition sites (FRT) or loxP sites allowing the excision by the site-specific recombinases Flp and Cre, respectively, some remnants of the recombination remain in the mutated genome.

The great advantage of RecE/T cloning is that the linear sequence containing a selectable marker flanked by homologous sequences can be provided by PCR amplification with synthetic oligonucleotide primers, and no cloning is required. These features make it a powerful tool for gene deletions in MDV BACs, and more than 30 mutant viruses have been generated using this technique (Table 3.1).

The second method of mutagenesis for allelic exchange is RecA-mediated homologous recombination that requires cloning of a transfer plasmid. The desired mutation is cloned into a suicide plasmid and flanked by homologous viral sequences of approximately 500 to 1000 bp in length that are required for recombination (Figure 3.2). Allelic exchange occurs via a two-step process of co-integrate formation and resolution after the introduction of the cloned suicide plasmid into *E. coli* harbouring the BAC. The suicide plasmid is lost after resolution of the co-integrate as a result of its temperature-sensitive origin of replication. So far we have generated a number of U_L44 mutant viruses using this method, which is especially helpful when seamless modifications of the genomes are required – for example, when manipulations in highly sensitive areas like the repeat regions have to be performed, or when point mutations are desired (Figure 3.2). Briefly summarized, the pST76K_SR transfer vector that is routinely used for these manipulations harbours a recA gene, a kanamycin-resistance (kan^R) gene, a temperature-sensitive origin of replication, and the sacB gene (Figure 3.2). After introduction of the pST76K_SR containing the desired sequences into *E. coli*, the bacteria are grown at 42°C and co-integrates form by the action of RecA (Figure 3.2). The co-integrates are resolved by a second

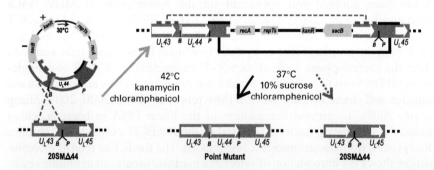

Figure 3.2 RecA-mediated mutagenesis of the U_L44 gene encoding gC. After cloning of mutated genes into pST76K_SR, the plasmids are electroporated into *E. coli* containing the gC-negative genome termed 20SMΔ44. RecA mediates recombination between the two circular molecules via the homologous sequences present in the transfer vector and the BAC. Plasmid pST76K_SR is cleared by growth at 42°C. Resulting co-integrates are resolved by plating on agar containing sucrose. This reaction is also RecA-mediated and results in either restoration of wild-type (dashed) or the formation of the desired mutant genome (black), which can readily be identified by the introduced restriction enzyme sites.

RecA-mediated reaction and in the presence of 10 per cent sucrose in the agar plates. The sacB gene present in all co-integrates confers sensitivity to sucrose, thereby facilitating the homologous recombination leading to the resolution of the co-integrates. Colonies harbouring resolved co-integrates are detected by replica plating on agar plates containing either kanamycin (kan^R is present in the co-integrates) or chloramphenicol (present in the mini-F plasmid sequences). The desired mutant genomes are finally identified and checked for correct insertion of the desired sequences by polymerase chain reaction (PCR), restriction enzyme analysis, Southern blotting, and nucleotide sequencing (Figure 3.2: Osterrieder *et al.*, 2003).

Summary

Until recently, MDV genome research was impeded by two major shortcomings: the paucity of sequence data, and the lack of fast, easy, and reliable methods to manipulate viral DNA. Both these roadblocks have been removed, and sequence analysis of MDV, GaHV-3 and HVT has confirmed that the three viruses are closely related but have evolved independently from each other, and thus should be considered as separate viruses and not as serotypes of one virus. The establishment of BAC and cosmid cloning (Reddy *et al.*, 2002) has greatly facilitated the generation and analysis of recombinant viruses, which will certainly speed discovery in the field. With the sequence information and the tools for MDV mutagenesis in hand, we should be able to move forward considerably in terms of understanding MDV genes and gene products and their functions in virus replication.

Acknowledgements

The authors thank Karsten Tischer, Daniel Schumacher and Sascha Trapp of Cornell College for Veterinary Medicine for critically reading the manuscript. Karsten Tischer provided Figure 3.2 and Daniel Schumacher Table 3.1.

References

Afonso, C. L., Tulman, E. R., Lu, Z. *et al.* (2001). *J. Virol.*, **75**, 971–978.
Anderson, A. S., Francesconi, A. and Morgan, R. W. (1992). *Virology*, **189**, 657–667.
Anderson, A. S., Parcells, M. S. and Morgan, R. W. (1998). *J. Virol.*, **72**, 2548–2553.
Aouacheria, A., Banyai, M., Rigal, D. *et al.* (2003). *Virology*, **316**, 256–266.
Baigent, S. J., Ross, L. J. N. and Davison, T. F. (1998). *J. Gen. Virol.*, **79**, 2795–2802.
Barrow, A. D., Burgess, S. C., Baigent, S. J. *et al.* (2003). *J. Gen. Virol.*, **84**, 2635–2645.

Bradley, G., Lancz, G., Tanaka, A. and Nonoyama, M. (1989). *J. Virol.*, **63**, 4129–4135.

Brune, W., Messerle, M. and Koszinowski, U. H. (2000). *Trends Genet.*, **16**, 254–259.

Calnek, B. W. (2001). In *Marek's Disease* (ed. K. Hirai), pp. 25–55. Springer-Verlag, Berlin.

Calnek, B. W., Schat, K. A., Ross, L. J. and Chen, C. L. (1984). *Intl J. Cancer*, **33**, 399–406.

Cantello, J. L., Anderson, A. S. and Morgan, R. W. (1994). *J. Virol.*, **68**, 6280–6290.

Cantello, J. L., Parcells, M. S., Anderson, A. S. and Morgan, R. W. (1997). *J. Virol.*, **71**, 1353–1361.

Cohen, J. I. and Nguyen, H. (1997). *J. Virol.*, **71**, 6913–6920.

Cohen, J. I. and Seidel, K. (1994). *J. Virol.*, **68**, 7850–7858.

Davison, A. (2002a). *Virus Res.*, **82**, 127–132.

Davison, A. J. (2002b). *Vet. Microbiol.*, **86**, 69–88.

Davison, A., Eberle, R., Desrosiers, R. C. *et al.* (2002a) *Virus taxonomy*. In: www.ncbi.nlm.nih.gov./ICTVdb/Ictv/fs_herpe.htm

Davison, A. J., Dargan, D. J. and Stow, N. D. (2002b). *Antiviral Res.*, **56**, 1–11.

del Rio, T., Werner, H. C. and Enquist, L. W. (2002). *J. Virol.*, **76**, 774–782.

Dorange, F., El Mehdaoui, S., Pichon, C. *et al.* (2000). *J. Gen.Virol.*, **81**, 2219–2230.

Dorange, F., Tischer, B. K., Vautherot, J. F. and Osterrieder, N. (2002). *J. Virol.*, **76**, 1959–1970.

Fragnet, L., Blasco, M. A., Klapper, W. and Rasschaert, D. (2003). *J. Virol.*, **77**, 5985–5996.

Gray, W. L., Baumann, R. P., Robertson, A. T. *et al.* (1987). *Virology*, **158**, 79–87.

Hayashi, M., Kawamura, T., Akaike, H. *et al.* (1999). *J. Vet.Med.Sci.*, **61**, 389–394.

Hong, Y. and Coussens, P. M. (1994). *J. Virol.*, **68**, 3593–3603.

Hong, Y., Frame, M. and Coussens, P. M. (1995). *Virology*, **206**, 695–700.

Izumiya, Y., Jang, H. K., Kashiwase, H. *et al.* (1998). *J. Gen. Virol.*, **79**, 1997–2001.

Izumiya, Y., Jang, H. K., Sugawara, M. *et al.* (1999). *J. Gen. Virol.*, **80**, 2417–2422.

Izumiya, Y., Jang, H. K., Ono, M. and Mikami, T. (2001). In *Marek's Disease* (ed. K. Hirai), pp. 191–221. Springer-Verlag, Berlin.

Jang, H. K., Ono, M., Kim, T. J. *et al.* (1998). *Virus Res.*, **58**, 137–147.

Kawamura, M., Hayashi, M., Furuichi, T. *et al.* (1991). *J. Gen. Virol.*, **72**, 1105–1111.

Kingham, B. F., Zelnik, V., Kopacek, J. *et al.* (2001). *J. Gen. Virol.*, **82**, 1123–1135.

Kopacek, J., Ross, L. J., Zelnik, V. and Pastorek, J. (1993). *Acta Virol.*, **37**, 191–195.

Lee, L. F., Wu, P., Sui, D. X. *et al.* (2000). *Proc. Natl Acad. Sci. USA*, **97**, 6091–6096.

Li, D. S., Pastorek, J., Zelnik, V. *et al.* (1994). *J. Gen. Virol.*, **75**, 1713–1722.

Lupiani, B., Lee, L. F. and Reddy, S. M. (2001). In *Marek's Disease* (ed. K. Hirai), pp. 159–190. Springer-Verlag, Berlin.

Mallory, S., Sommer, M. and Arvin, A. M. (1997). *J. Virol.*, **71**, 8279–8288.

Mallory, S., Sommer, M. and Arvin, A. M. (1998). *J. Infect. Dis.*, **178** (Suppl. 1), 22–26.

Maotani, K., Kanamori, A., Ikuta, K. *et al.* (1986). *J. Virol.*, **58**, 657–660.

Maray, T., Malkinson, M. and Becker, Y. (1988). *Virus Genes*, **2**, 49–68.

McKie, E. A., Ubukata, E., Hasegawa, S. *et al.* (1995). *J. Virol.*, **69**, 1310–1314.

Messerle, M., Crnkovic, I., Hammerschmidt *et al.* (1997). *Proc. Natl Acad. Sci. USA*, **94**, 14759–14763.

Morgan, R. W., Xie, Q., Cantello, J. L. *et al.* (2001). In *Marek's Disease* (ed. K. Hirai), pp. 223–243. Springer-Verlag, Berlin.

Muyrers, J. P., Zhang, Y., Testa, G. and Stewart, A. F. (1999). *Nucleic Acids Res.*, **27**, 1555–1557.

Muyrers, J. P., Zhang, Y., Buchholz, F. and Stewart, A. F. (2000). *Genes Dev.*, **14**, 1971–1982.

Muyrers, J. P., Zhang, Y. and Stewart, A. F. (2001). *Trends Biochem. Sci.*, **26**, 325–331.

Osterrieder, N., Schumacher, D., Trapp, S. *et al.* (2003). *Berl. Münch. Tierarztl Wochenschr.*, **116**, 373–380.

Parcells, M. S., Anderson, A. S. and Morgan, R. W. (1994). *Virus Genes*, **9**, 5–13.

Parcells, M. S., Dienglewicz, R. L., Anderson, A. S. and Morgan, R. W. (1999). *J. Virol.*, **73**, 1362–1373.

Parcells, M. S., Lin, S. F., Dienglewicz, R. L. *et al.* (2001). *J. Virol.*, **75**, 5159–5173.

Parcells, M. S., Arumugaswami, V., Prigge, J. T. *et al.* (2003). *Poultry Sci.*, **82**, 893–898.

Peng, F. Y., Specter, S., Tanaka, A. and Nonoyama, M. (1994). *Intl J. Oncol.*, **4**, 799–802.

Petherbridge, L., Howes, K., Baigent, S. *et al.* (2003). *J. Virol.*, **77**, 8712–8718.

Pratt, W. D., Cantello, J., Morgan, R. W. and Schat, K. A. (1994). *Virology*, **201**, 132–136.

Preston, C. M. (2000). *J. Gen. Virol.*, **81**, 1–19.

Qian, Z., Brunovskis, P., Rauscher, F. *et al.* (1995). *J. Virol.*, **69**, 4037–4044.

Reddy, S. M., Lupiani, B., Gimeno, I. M. *et al.* (2002). *Proc. Natl Acad. Sci. USA*, **99**, 7054–7059.

Ren, D. L., Lee, L. F. and Coussens, P. M. (1994). *Virology*, **204**, 242–250.

Robertson, A. T., Caughman, G. B., Gray, W. L. *et al.* (1988). *Virology*, **166**, 451–462.

Roizman, B. (1982). In *The Herpesviruses* (ed. B. Roizman), pp. 1–23. Plenum Press, London.

Roizman, B. (1996). In *Fields Virology* (ed. B. N. Fields, D. M. Knipe and P.M. Howley), pp. 2221–2230. Lippincott-Raven Press, New York.

Roizman, B. (1999). *Acta Virol.*, **43**, 75–80.

Roizman, B. and Sears, A. E. (1996). In *Fields Virology* (ed. B. N. Fields, D. M. Knipe and P. M. Howley), pp. 2231–2295. Lippincott-Raven Press, New York.

Ross, N. L. J. (1999). *Trends Microbiol.*, **7**, 22–29.

Ross, N., Binns, M. M., Sanderson, M. and Schat, K. A. (1993). *Virus Genes*, **7**, 33–51.

Ross, N., O'Sullivan, G., Rothwell, C. *et al.* (1997). *J. Gen. Virol.*, **78**, 2191–2198.

Sakaguchi, M., Urakawa, T., Hirayama, Y. *et al.* (1993). *Virology*, **195**, 140–148.

Schat, K. A., Buckmaster, A. and Ross, L. J. (1989). *Intl J. Cancer*, **44**, 101–109.

Schumacher, D., Tischer, B. K., Fuchs, W. and Osterrieder, N. (2000). *J. Virol.*, **74**, 11088–11098.

Schumacher, D., Tischer, B. K., Reddy, S. M. and Osterrieder, N. (2001). *J. Virol.*, **75**, 11307–11318.

Shigekane, B., Kawaguchi, Y., Shirakata, M. *et al.* (1999). *Arch. Virol.*, **144**, 1893–1907.

Silva, R. F., Reddy, S. M. and Lupiani, B. (2003). *J. Virol.*, **78**, 733–740.

Smith, R. H., Caughman, G. B. and O'Callaghan, D. J. (1992). *J. Virol.*, **66**, 936–945.

Smith, R. H., Holden, V. R. and O'Callaghan, D. J. (1995). *J. Virol.*, **69**, 3857–3862.

Tischer, B. K., Schumacher, D., Beer, M. *et al.* (2002a). *J. Gen. Virol.*, **83**, 2367–2376.

Tischer, B. K., Schumacher, D., Messerle, M. *et al.* (2002b). J. Gen. Virol., **83**, 997–1003.

Tsushima, Y., Jang, H. K., Izumiya, Y. *et al.* (1999). *Virus Res.*, **60**, 101–110.

Tulman, E. R., Afonso, C. L., Lu, Z. *et al.* (2000). *J. Virol.*, **74**, 7980–7988.

Ui, M., Endoh, D., Cho, K. O. *et al.* (1998). *J. Vet. Med. Sci.*, **60**, 823–829.

Xie, Q., Anderson, A. S. and Morgan, R. W. (1996). *J. Virol.*, **70**, 1125–1131.

Xing, Z., Xie, Q., Morgan, R. W. and Schat, K. A. (1999). *Acta Virol.*, **43**, 113–120.

Zhang, Y., Muyrers, J. P., Testa, G. and Stewart, A. F. (2000). *Natl Biotechnol.*, **18**, 1314–1317.

Zhu, G. S., Iwata, A., Gong, M. *et al.* (1994). *Virology*, **200**, 816–820.

Marek's disease virus oncogenicity: molecular mechanisms

4

VENUGOPAL NAIR* and **HSING-JIEN KUNG****

* Institute for Animal Health, Compton Laboratory, Newbury, Berkshire, UK
** UC Davis Cancer Center, UCDMC, Sacramento, California, USA

Introduction

Marek's disease virus (MDV) is one of the most potent oncogenic herpesvirus known. Classified as an α-herpesvirus according to DNA sequence homology and genome organization (Lee *et al.*, 2000a; Tulman *et al.*, 2000), the biological properties of MDV are more akin to those of γ-herpesviruses such as Epstein-Barr virus (EBV), herpesvirus saimiri (HVS) and human herpesvirus 8 (HHV-8), also referred to as Kaposi's sarcoma herpesvirus (KSHV). The studies of EBV, HVS and HHV-8 have provided an impressive amount of information concerning herpesvirus oncogenesis (Damania and Jung, 2001). Similar to these viruses, MDV establishes latency in lymphoid cells, some of which are subsequently transformed. Latency is established primarily in activated or semi-activated T cells, which then spread to peripheral sites via the circulation and leave blood vessels and reactivate virus at peripheral sites, inducing a secondary cytolytic infection in various epithelial and secondary lymphoid tissues (Calnek *et al.*, 1984). Tumours induced by MDV are lymphomas comprised of mixed lymphocytes and monocytes, the transformed elements of these masses being primarily CD4$^+$ T cells (Nazerian and Sharma, 1975). Cell lines established from MDV-induced lymphomas are predominantly CD4$^+$ T cells (Nazerian and Sharma, 1975; Schat *et al.*, 1982, 1991; Parcells *et al.*, 1999). Thus, there is an apparent link between latency and transformation.

The onset of MDV-induced tumours is very rapid, sometimes as early as 3 weeks. The ability to induce rapid-onset tumours, resembling those caused by some of the acutely transforming retroviruses, suggests the direct involvement of virus-encoded transforming gene(s) in oncogenesis. In the past decade, several approaches have been taken by investigators to identify potential oncogenes

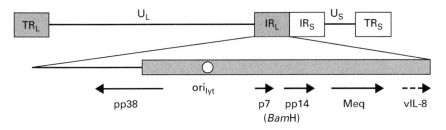

Figure 4.1   Structure of the MDV genome. The genome contains two unique regions, the U_L and the U_S regions, which are flanked respectively by the repeats TR_L and IR_L and TR_S and IR_S regions (see Chapter 3 for detailed structure of the MDV genome). The enlarged IR_L region and the major open reading frames are indicated.

of MDV. The first was to use comparative genomics to identify genes unique to the onco-serotypes of MDV (i.e. present only in MDV-1, and absent in non-pathogenic serotype MDV-2 and the herpesvirus of turkey (HVT) strains). Such genome-wide sequence comparison became possible with the availability of the complete genome sequences for all three serotypes. The MDV family has a genomic organization very similar to herpes simplex virus (HSV), with almost one-to-one correspondence of genes encoded in the U_L and U_S regions. The repeat (R) regions, however, are unique among MDV serotypes, as well as very divergent from HSV (see Chapter 3). Further approaches included the identification of regions grossly altered during attenuation of MDV, and the characterization of genes that are expressed in transformed cells. All these approaches converged to implicate the R_L (repeats flanking the U_L) region in oncogenesis (Figure 4.1). Some of the R_L-encoded genes, which are unique to MDV, are Meq (Jones *et al.*, 1992; Peng *et al.*, 1995), vIL-8 (Parcells *et al.*, 2001), pp38 (Cui *et al.*, 1990) and two small open reading frames (ORF), pp14 and p7 (Hong and Coussens, 1994).

Meq, a Jun/Fos family member, as a putative oncogene for MDV

Among these genes, only Meq is the most consistently expressed latency gene (Parcells *et al.*, 2001). As will be described later, Meq encodes a gene resembling a fusion of the oncogene Jun and the tumour suppressor gene WT-1 (Wilm's tumour-1). Accumulating evidence based on its transforming properties and studies on deletion mutants suggest that Meq is likely to be the principal oncogene for MDV, with other MDV genes serving auxiliary functions. This review focuses on the biochemical and oncogenic properties of Meq in the context of MDV oncogenesis. A brief overview of the roles of other MDV genes potentially associated with latency/transformation is also provided at the end of the chapter.

The structure and function of Meq

Structure

Meq is a 339 amino-acid protein, characterized by an N-terminal bZIP domain and a proline-rich C-terminal transactivation domain (Jones and Kung, 1992; Jones *et al.*, 1992). Meq is the only herpesviral bZIP protein within the immediate family of Jun/Fos. The C-terminal domain contains two and one-half repeats of proline-rich sequences with several PPPP and PXXP motifs, known protein–protein interaction modules (especially for SH3-containing proteins). In addition to these two major domains, there is an additional basic region (BR1) at the N-terminus. For this reason, the basic region in the bZIP domain is called BR2. Meq is localized primarily in the nucleus, including the nucleoplasm, nucleolus and coiled bodies. BR1 and BR2 are both nuclear localization signals (Liu *et al.*, 1997). BR2, highly rich in arginine, is also responsible for Meq's nucleolar localization. The biological significance of the different localizations of Meq is presently unclear, although it has been shown that Meq interacts with, and is phosphorylated by, CDK2 in coiled bodies (Liu *et al.*, 1997, 1998). The various subcellular locations allow Meq to interact with different cellular factors and perform multiple functions – a common trait of viral regulators.

Transactivation

The entire C-terminal domain of Meq (137 to 339 aa), when linked to the Gal4 DNA-binding domain, can strongly activate Gal4-dependent target genes, indicating that it is a *bona fide* transactivation domain (Qian *et al.*, 1996). Askovic and Baumann (1997) extended this finding by showing that this domain has a transcriptional potency comparable to that of HSV VP16, when fused to the EBV Zta DNA-binding domain (Askovic and Baumann, 1997). This transactivation ability depends on the presence of both the proline-rich domain and the C-terminal 33 amino acids. Interestingly, if only the proline-rich repeats are used in a similar transactivation analysis, they repress, rather than activate, transcription (Qian *et al.*, 1996). This is reminiscent of the proline-rich domain of WT-1 tumour suppressor, which is strongly suppressive (Call *et al.*, 1990). We speculate that Meq can function both as a transactivator and as a repressor, depending on whether the proline-rich repeats are exposed. This in turn depends on the phosphorylation state of Meq, the dimerization partner and the promoter context. Indeed, it was recently shown that wild-type Meq transactivates its own promoter but represses the transcription of the pp38/pp14 divergent promoter and of the ICP4 promoter (Liu *et al.*, 1999a; Liu and Kung, 2000). As shown below, MDV strains with varying degrees of virulence bear clustered mutations or insertions in the proline-rich repeats, implicating this region in the pathogenesis of the virus. Remarkably, it was shown that more virulent strains, which carry fewer proline-rich repeats or have mutations at the proline residues, have higher activation and lower repression potential, consistent with the above hypothesis (M. S. Parcells, University of Arkansas, personal communication).

Dimerization in vitro and in vivo

There are four major classes of bZIP proteins in mammals: Jun (c-Jun, JunB and JunD), Fos (Fos, FosB and Fra), ATF (CREB, ATF1 to 6) and C/EBP (α and β). In chickens, only c-Jun, JunD, Fos Fra, and ATF2 have been identified. The ZIP domain is known to form a dimer with another ZIP domain, allowing the adjacent basic region to anchor to DNA. The pairing of the dimer is via the regularly-spaced leucines, but the specificity and affinity depend on the charges born by the amino acids surrounding the leucine residues (Kung *et al.*, 2001a). The interaction is favoured if the charges are neutralized in the leucine zipper coiled-coil structure (Schuermann *et al.*, 1991; Baxevanis and Vinson, 1993; Glover and Harrison, 1995; Liu and Kung, 2000; Kung *et al.*, 2001b). We found, both experimentally and based on the 'charge rule', that Meq has multiple dimerization partners. It binds to itself, c-Jun, JunB, ATF2 and Fos with a high affinity (Brunovskis *et al.*, 1996), but much less so with CREB, ATF1, ATF3 and C/EBP. The most stable dimer formed by far is the Meq/c-Jun heterodimer, such that if Meq and Jun are mixed in equimolar concentrations, nearly all of the molecules are Meq/Jun heterodimers (Qian *et al.*, 1996). MSB-1, a Meq-expressing MDV-transformed T-cell line harbouring a latent MDV genome, expresses c-Jun and Fra, but not Fos or JunD. In this cell line, c-Jun is found to be the principal dimerization partner of Meq by the following criteria. Meq is co-localized with c-Jun in the nucleus, co-precipitated with c-Jun in nuclear extracts and co-recruited to promoters containing AP-1 sites (see below) such as those of chicken interleukin 2 (IL-2) and Meq itself (Levy *et al.*, 2003). These results suggest that in chicken T cells, Meq may function through the Jun pathway. It also indicates that Meq can activate the expression of IL-2, an important T-cell growth factor, and autoregulate its own expression during latency.

DNA and chromosomal binding

The basic region of bZIP proteins dictates its DNA-binding specificity, although the dimerization partner can modulate its affinity. We used a PCR-based, unbiased approach, CASTing (Cyclic Amplification of Selected Targets), to define the DNA-binding sequences of Meq, which were then verified by footprint analysis (Qian *et al.*, 1996). Meq/Jun heterodimer binds both AP-1 and CRE sites. One such site is present in the promoter of Meq, suggesting autoregulation of its expression. Meq/Meq homodimer also selects the AP-1 site, but it is a low-affinity binding site. The high-affinity binding site for the Meq/Meq homodimer has a core sequence of ACACA. One such site is at the MDV origin of replication, coincident with the divergent promoter of pp38 and pp14. Recent experiments with chromosomal immunoprecipitation techniques (Chen *et al.*, 1997, 1999; Louie *et al.*, 2003), using the MDV latent genome in

MSB-1 as the natural chromosomal targets, have demonstrated that these predicted sites are where Meq binds *in vivo* (Levy *et al.*, 2003). It was found that Meq interactions are enriched in three regions of the latent MDV chromosome – the Meq promoter, the ICP4 promoter and the MDV origin of replication (or the pp14/pp38 promoters) respectively. It was found that both Meq and Jun are recruited to the Meq promoter, which contains the AP-1 site, but only Meq is recruited to the pp14/pp38 promoter, which contains the ACACA sequence. Reporter assays showed that Meq/Jun transactivates the Meq promoter, but not the pp14/pp38 promoter, whereas Meq represses the pp14/pp38 promoter. These studies showed that, depending on the dimerization partner and target sites, Meq exerts opposing biological effects. The data also raised an interesting issue concerning Meq's role in latency and replication.

Interactions with cellular factors

In addition to its dimerization partners, Meq is associated with a number of cellular factors involved in cell cycle regulation, notably p53 and RB (Brunovskis *et al.*, 1996). The binding between Meq and p53 involves the bZIP domain of Meq and the C-terminal tetradimerization domain of p53. The binding domain of Meq to RB is also mapped to the N-terminal domain of Meq, where there is a putative RB-binding motif, LXCXE, at the end of the zipper region. Whether these residues are involved in the direct binding with RB, and the significance of RB binding, remain to be established. Recent data also indicate interactions between Meq and CtBP (C-terminal binding protein), a highly conserved cellular protein that functions as a transcriptional co-repressor (Hickabottom *et al.*, 2002). Originally identified through its interaction with the C-terminus of the adenovirus oncoprotein E1A, CtBP has been shown to antagonize the transcriptional activation function of E1A. Whether the interaction of Meq and CtBP has any role in modulating the oncogenic potential of Meq remains to be demonstrated. At the cellular level, Meq is co-localized with PCNA, PML and CDK2. The co-localization with CDK2 in Cajal (formerly coiled) bodies was particularly interesting (Liu *et al.*, 1999b), leading to the subsequent discovery that CDK2 functions in the transcription of histone in Cajal bodies (Liu *et al.*, 2000). In Cajal bodies CDK2 is in the activated form, presumably due to the phosphorylation by CAK. In Meq-transformed cells, there is a considerably higher concentration of CDK2 in the Cajal bodies. This, coupled with Meq's binding to p53 and RB, presumably results in the deregulation of the cell cycle checkpoint as observed in the transformed cells. The co-localization of CDK2 with Meq has an added significance, as CDK2, but not other protein kinases tested, phosphorylates Meq at serine residue 42, and translocates it into the cytoplasm as a means of regulating its nuclear activity (Liu *et al.*, 1999b). Indeed, Meq is partially localized in the cytoplasm in the S phase when CDK2 activity is the highest. A Meq mutant with S changed to D, mimicking the negative charge of the phosphorylated form, caused Meq to be located in the cytoplasm.

This mutant will be valuable for studying functions of Meq in the cytoplasm. In this regard Meq is phosphorylated by a number of cytoplasmic kinases, including PKA, PKC and MAPK, and the proline-rich domain is able to bind the SH3 domain of Src. Much work is required to understand the significance of the post-translational modifications of Meq.

Meq as a transcriptional regulator

Based on the biochemical properties of Meq, it was postulated that Meq could synergize with c-Jun and transcribe genes targeting AP-1 sites. Recent micro-array analysis of genes differentially regulated in Meq-transformed versus wild-type DF-1 cells generally supports this hypothesis, although it is also clear that Meq has non-Jun related functions. This was made possible by the availability of a newly-developed chicken cDNA microarray (Neiman *et al.*, 2001) which contains approximately 3400 confirmed chicken expressed sequence tag (EST) sequences, derived from a mixture of cDNA libraries including those from B cell-derived DT-40 cells, T cells and other mixed tissues. Genes with statistically significant changes were identified and those related to oncogenesis were selected and confirmed by RT-PCR (A. Levy and H. -J. Kung, unpublished observations). These differentially regulated genes fall into four major categories:

1. *The survival pathways*: Bcl2, Bcl9 and Ski (the latter a proto-oncogene that interacts with Smads and intercepts the transforming growth factor β (TGF-β) pathway (Stavnezer *et al.*, 1981, 1989)) are up-regulated, whereas FasL, Fas and DAP5 are all down-regulated, consistent with Meq's ability to confer anti-apoptotic properties.
2. *Oncogenes*: c-Jun, c-Myc, and c-Ski are activated.
3. *Transformation-related target genes of Jun*: JAC, JTAP1, HBEGF (Hadman *et al.*, 1998; Fu *et al.*, 1999; Bader *et al.*, 2000; Tusher *et al.*, 2001) are up-regulated.
4. *Immunoregulatory genes*: IL-8, RANTES, MIP and TGFβ are up-regulated.

The above data suggest that Meq has versatile roles in oncogenesis, and the major oncogenic pathway it engages is likely to be that of Jun and Ski. Indeed, Jun and Ski work in concert to effect anti-apoptosis (Pessah *et al.*, 2002). It is noteworthy that the genes up-regulated by Meq are more similar to those up-regulated by v-Jun (the oncogene carried by retroviruses) than by its cellular homologue c-Jun. v-Jun, due to its deletion of a domain targeted for ubiquitination, is more stable than c-Jun (Treier *et al.*, 1994). Dimerization of Meq with c-Jun also affects the stability of c-Jun, rendering it more like v-Jun. Meq has additional structural features more akin to v-Jun than c-Jun. For instance, one of the activating mutations of v-Jun is the change of serine 226 in c-Jun to phenylalanine in v-Jun. Meq carries a phenylalanine in this position. Thus, to some extent, Meq/Jun heterodimers may have some of the transforming properties of v-Jun. It is also interesting that Meq induces chemokines and cytokines to modulate immune cells. Like lymphoma cells, Meq-transformed cells release TGFβ as a means of suppressing surrounding immune cells. The activation of Ski in these cells is one way of allowing them to escape self-imposed growth arrest and apoptosis.

Meq as a transforming protein

Because of the lack of a proper chicken T-cell transformation system, the transforming properties of Meq were first studied on rodent fibroblast cell lines such as Rat-2 cells. Overexpression of Meq by infection of cells with a retrovirus vector carrying Meq led to serum-independent and anchorage-independent growth, which was accompanied by striking morphological changes and a shortened G1 phase (Liu *et al.*, 1997). On the other hand, if Meq was introduced by DNA-mediated transfection, the expression level was lower and a complementary oncogene such as ras was required to achieve full transformation phenotype. This behaviour is similar to v-Jun or overexpressed c-Jun, which also complements ras in this type of transformation assay. The chimaeric molecule Meq (bZIP)-Jun (TA), which fuses the bZIP domain of Meq and the TA (transactivation) domain of Jun, is as potent as wild-type Meq or Jun in this assay. Likewise, the reciprocal construct Meq (TA)-Jun (bZIP) also exhibits transforming properties. These results, together with the finding that Meq and c-Jun co-localize in the transformed cells, provide strong evidence that Meq and c-Jun participate in similar signal transduction pathways, and that the different domains of these molecules are interchangeable with respect to this particular function (Liu *et al.*, 1999b).

In addition to the transforming and mitogenic properties of Meq, Meq-infected cells are highly resistant to apoptosis induced by serum withdrawal, tumour necrosis factor (TNF)α, UV-irradiation and C2-ceramide (Liu *et al.*, 1998). Thus Meq seems to induce a general protective pathway, and, given its transactivation ability, a plausible mechanism is that it mediates transcriptional regulation of apoptosis-related molecules such as Bcl2 and Bax, as shown in the Meq-Rat-2 cell transformation model (Liu *et al.*, 1997). In addition, Meq-transformed cells have a shortened G1 phase, consistent with the deregulation of the cell cycle checkpoint by activated CDK2 (Liu *et al.*, 1997).

These studies on Meq-mediated transformation were recently extended to DF-1, a spontaneously immortalized chicken cell line (Himly *et al.*, 1998). This cell line does not exhibit any transformed phenotypes *in vitro* and tumorigenicity *in vivo*. DF-1 cells have been used for studying transformation by a number of avian retroviral oncogenes such as v-ErbB, v-Src, v-Jun, v-Myc, v-Fos, v-Ski, as well as weakly transforming c-Jun (Himly *et al.*, 1998). The cell lines exhibit a different transformed morphology with different oncogenes, providing a unique opportunity to study the cell biology and the signal pathways (Kim *et al.*, 2001a, 2001b). When Meq was transfected into DF-1 cells and stable clones isolated, the cells displayed radiant-shaped phenotypes, resembling cells transformed by the Ski oncogene, and formed large anchorage-independent colonies in soft agar. In addition, the Meq-transformed DF1 cells showed increased potential for serum-independent survival. All these properties echoed those found with Meq-Rat-2 cells (Liu *et al.*, 1997), and reinforce the notion that Meq is transforming. Most importantly, these studies showed, for the first time, the transforming potential of Meq in chicken cells.

These results show that Meq has transforming and anti-apoptotic properties when introduced into immortalized cells. A key question is whether Meq alone can transform primary cells. Attempts have been made to place Meq in a replication-competent avian retrovirus vector RCAS (Petropoulos and Hughes, 1991) and use it to infect chicken embryo fibroblasts (CEF) or bone marrow cultures *in vitro* or chickens *in vivo*. Most of these studies gave negative results except for subtle morphological changes in CEF, similar to those induced by c-Jun expressing viruses (Bos *et al.*, 1990; Petropoulos and Hughes, 1991; Morgan *et al.*, 1994). In retrospect, this is not unexpected, as most herpesviral oncogenes are not as potent as the retroviral oncogenes. We suggest Meq is weakly oncogenic for primary cells and requires co-operation of additional viral and host genes to cause full malignancies involving target lymphocytes.

Meq as a major oncoprotein in MDV oncogenesis

Perhaps the strongest evidence that Meq is an MDV oncogene came from Meq knockout experiments. Using antisense Meq, Xie *et al.* (1996) demonstrated that down-regulation of Meq resulted in the loss of the colony formation ability of MSB1, an MDV-transformed tumour cell line. Dr R. W. Morgan's laboratory (University of Delaware), using the marker rescue approach (Anderson *et al.*, 1998), and Dr S. M. Reddy's laboratory (ADOL, East Lansing), by rescue from overlapping cosmid library (unpublished observations), both report the successful development of recombinant MDV lacking Meq. In both cases, the Meq-null mutant viruses replicated well *in vitro* but had completely lost their *in vivo* oncogenicity (Table 4.1). The Meq knockout virus replicated like wild-type

Table 4.1 Incidence of MD in birds inoculated with different recombinant viruses.

Virus	MD incidence	PBL titres at 2 weeks
ΔvIL-8[1]	9% ($n = 23$), 14% ($n = 7$)	1–2% of wt
Δpp38[2]	27% ($n = 40$)	0.8% of wt
ΔMeq[3]	0% ($n = 17$)	2% of wt

wt, wild-type virus; n = number of birds used.
[1]Parcells *et al.* (2000) RB1B-based mutant.
[2]Reddy *et al.* (2001) Md5-based mutant.
[3]S. M. Reddy (personal communication) Md5-based mutant.

virus *in vivo* during the first week in the primary cytolytic infection. However, the titres decreased to about 2 per cent of those for the wild-type virus during the second week, coinciding with entry into latency and the reactivation phase. This suggested that Meq might play some role in latency or reactivation, since Meq-null mutant was able to establish latency in the T-cell model (see below).

The lower titres in the second cytolytic infection undoubtedly may contribute to the lower oncogenicity. However, alone it does not account for the complete lack of oncogenicity of the Meq knockout mutant, as both pp38-knockout (Reddy *et al.*, 2002) and vIL-8 knockout (Parcells *et al.*, 2001), which have comparable or even lower titres (0. 8 per cent and 1. 0 per cent of the wild type at two weeks, respectively), still retain significant oncogenicity (27 per cent and 10 per cent MD incidence, respectively). Tumour cell lines derived using pp38 and vIL-8 knockout viruses, which harbour the latent defective viral genomes, have been produced. This is, however, not the case for Meq, suggesting that it plays a critical role in the transformation process.

Meq mutations in proline-rich domains correlate with virulence

If Meq is the principal oncogene of MDV, one might expect to find mutations in the Meq gene among MDV isolates with varied oncogenicity or pathogenicity. Depending on the virulence index (lymphoma incidence, severity of immune-suppression and nerve lesions etc.), MDV has been classified as vv+MDV (very virulent+MDV), vvMDV, vMDV pathotypes (Witter, 1997) and the nv (non-virulent) MDV vaccine strains. The vv+MDV pathotypes were isolated mostly in the 1990s from chickens vaccinated with a bivalent vaccine containing HVT and MDV-2 (Witter, 1997), and the vvMDV strains in the 1980s from HVT-vaccinated birds (Witter, 1983). The Meq loci of isolates representing the different pathotypes of MDV were recently sequenced (M. S. Parcells, University of Arkansas, personal communication). While there is no strict correlation between the genotypes and the pathotypes, an interesting picture has emerged. The mutations were mostly in the proline-rich region. The Meq sequence of the vMDV (GA strain) and vvMDV (RB1B strain) pathotypes showed multiple PPPP repeats in the proline-rich region. L-Meq (large Meq) with an in-frame insertion of 59 amino acids of the proline-rich repeat was found in an MDV-1 vaccine strain (CV1988) and in the low virulence JM-10 and MKT strains (Figure 4.2). Interestingly, all vv+MDV strains tested (CD, MK, RL, TK and U) carried mutations in PPPP repeats. While it is hard to rule out that these mutations are coincidental, we note that these viruses were isolated from geographically separated flocks of different ages (broilers, broiler-breeders and layer flocks) throughout the USA. L-Meq was first discovered by Dr M. Onuma's laboratory (Lee *et al.*, 2000b). It was recently shown in a transactivation assay that L-Meq is 'super-repressive' (Chang *et al.*, 2002), consistent with the notion that the transactivation potential of Meq is the basis for its oncogenicity.

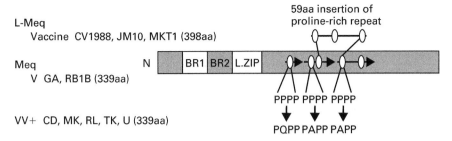

Figure 4.2   Natural variants of the Meq protein. Wild-type Meq, represented by the GA strain Meq (339 aa), carries two and one-half of the proline-rich repeats (arrows) with multiple PPPP motifs (oval shape). L-Meq, represented by the vaccine strain CVI988, has a 59 aa proline-rich repeat insert (top). The very virulent plus (vv+MDV) represented by the CD, MK, RL, TK and U strains carry mutations at the second position of the PPPP motif.

Meq's potential roles in MDV replication and latency

The data presented above suggest that Meq is equipped with transforming potential, yet it is most likely that Meq was initially acquired by MDV-1 for the purpose of replication and latency. Indeed, most of the oncogenes of DNA tumour viruses, such as the T antigen for SV40, E1A and E1B for adenovirus, encode replicating enzymes that acquire additional roles in transformation. Meq is predominantly expressed in the latent state, although it has been reported that Meq may also be expressed early in infection (Jones *et al.*, 1992). Since latency is a prerequisite of transformation or tumorigenesis, Meq's role in latency may be as important as its transforming properties in tumorigenesis. The Meq knockout mutant displays a replication property *in vitro* in CEF, no different from the wild type. The *in vivo* replication capacity of the Meq knockout mutant, however, is significantly attenuated and the reduction in growth *in vivo* coincided with the time of entry to latency or reactivation from latency. This suggests that Meq may play a role in the latency process. The biochemical data reviewed earlier indicate that Meq binds to the Meq promoter, ICP4 promoter and the replication origin (pp14/pp38 divergent promoter). It is conceivable that Meq may function in suppressing the expression of latency switch genes, while maintaining its own expression. Recently, it was shown the ectopic expression of ICP4 can lead to the reactivation of MDV lytic replication in a latent cell line (V. Zelnik, Cuxhaven, unpublished data). Meq, by controlling the expression of ICP4, could be involved in latency maintainence or reactivation. A second way Meq may control latency is related to its binding at a site close to where the origin binding protein UL9 binds. It is conceivable that Meq binding may alter the chromatin structure or interfere with UL9 binding, such that the DNA polymerase is no longer recruited to this region. Alternatively, Meq may function in reactivation of the latent genome – a notion that seems to be contradictory to the observation of the persistent expression of Meq in the latency. Accumulating evidence suggests that Meq has several spliced forms, is modified by multiple kinases and has multiple partners. It is not difficult to imagine that these structural modifications could change a repressor to an activator, or *vice versa*.

Summary and implications of Meq as an oncogene for MDV

In summary, both biochemical and genetic studies carried out in recent years suggest that Meq is the principal oncogene of MDV. We have learned that Meq is a versatile protein that binds DNA, RNA and a variety of proteins. Meq has multiple cell cycle-dependent subcellular localizations, enabling it to interact with different cell signal molecules. With respect to transformation and oncogenesis, Meq's dimerization with Jun, Fos and ATF, and its binding to AP-1 sites, are likely to be critical in the transformation processes. AP-1 activity has long been recognized as a key component in T-cell transformation and activation (see below). Activated AP-1 is able to reprogramme the genomic expression, leading to a transformation phenotype. Meq, via dimerization with c-Jun, induces the expression of JTAP, JAC, HB-EGF, which themselves can propagate the transformation phenotypes. Meq was also found to activate the ski and myc oncogenes, known to be capable of inducing serum- and anchorage-independent growth. In addition, Meq interacts with CDK2, p53 and RB to effect cell cycle progression by shortening the G1 phase. Finally, Meq's ability to activate AP-1 activity and neutralize p53 activity may contribute to its strong anti-apoptotic effects. Indeed, in Meq-overexpressing cells anti-apoptotic genes such as Bcl2 are up-regulated and pro-apoptotic genes such as Bax, Fas, FasL and DAP5 are down-regulated (Figure 4.3). The results of the combined effects of serum- and anchorage-independent growth, the deregulation of the cell cycle and anti-apoptosis effects generate transformation phenotypes.

The studies summarized above were carried out primarily *in vitro*, and clearly they overlook the immune components involved in the *in vivo* tumorigenesis. In this regard, it is interesting to note that Meq also affects the expression of a number of immunomodulatory genes such as IL-8, RANTES and TGFβ. In addition, Meq might be involved in the modulation of the expression of CD30 antigen (see Chapter 8), which may have a profound effect on the tumorigenic properties of the transformed T cells.

It is noteworthy that the mechanisms proposed for Meq-mediated transformation have a close resemblance to those used by HTLV Tax in the transformation of T cells. The T-cell lymphoma induced by HTLV-1 involves mostly CD4$^+$ T cells, the same cell type targeted by MDV for transformation. In HTLV-transformed cell lines, AP-1 activities are generally high (Mori *et al.*, 2000; Iwai *et al.*, 2001). The oncogene of HTLV-1 is Tax, which itself is not a bZIP protein, but can directly bind CREB (Zhao and Giam, 1992), and catalyses

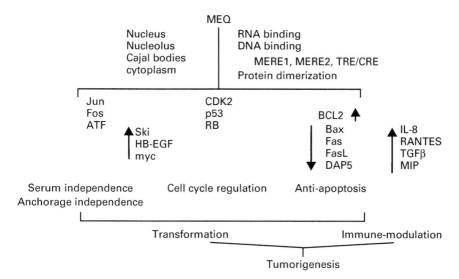

Figure 4.3 Predicted pathways of Meq-induced oncogenesis.

dimer formation between Jun/Fos and Jun/ATF family members (Wagner and Green, 1993). The JunD/Fra dimer is prevalent in the HTLV-1 transformed T cells (More *et al.*, 2000; Iwai *et al.*, 2001), and IL-2, a gene activated by AP-1, is often expressed to establish an autocrine loop (Mori *et al.*, 2002). Other properties associated with Tax expression include the enhanced expression of cyclins, the reduced expression of p21, the accelerated G1 to S transition (Liang *et al.*, 2002), the release of TGFβ and the inhibition of TGFβ-mediated transcriptional repression (Lee *et al.*, 2002). Several of these properties are shared by Meq expression. It thus appears that two different viruses from two different host species with a common type of target T cell may share similar transformation processes.

Other genes associated with pathogenicity

Various investigations aimed at unravelling the oncogenic determinants of MDV have led to the identification of a number of other genes that have potential roles in pathogenesis. As in the case for Meq, most of these genes also mapped to the transcriptionally active repeat (R_L/R_S) regions of the genome. Some of these MDV genes with a potential role in MD pathogenicity include: the *Bam*HI-H family of transcripts, the pp38/pp24 group of cytoplasmic phosphoproteins, the MDV-encoded CXC chemokine (vIL-8), the MDV-encoded telomerase RNA (vTR) subunit (Fragnet *et al.*, 2003) and the MDV ICP4 homologue-related transcripts.

*Bam*HI-H family of transcripts

The MDV *Bam*HI-H region received attention as a pathogenic determinant following demonstration of the expansion of the 132-bp tandem repeats within the *Bam*HI-D/*Bam*HI-H region during attenuation. The *Bam*HI-H region is transcriptionally active, and produces at least three transcripts (sizes 1.8 kB, 3.0 kB and 3.8 kB, respectively) derived from a bidirectional promoter shared with the pp38 gene. Several open reading frames (ORFs) have been predicted in the sequence of these transcripts (Peng *et al.*, 1992); however, protein products from only two of these transcripts have been reported. One of these, the ORF-A, encoded a 7-kDa polypeptide, designated BHa, that could be detected in CEF infected with oncogenic MDV strains as well as lymphoblastoid cell lines (Peng *et al.*, 1995). Similarly, a 14-kDa phosphoprotein (pp14) from another ORF (L-ORF10, MDV006/MDV075) was detected in the lysates of cells infected with serotype 1 MDV as well as in MSB-1 lymphoblastoid cells (Hong and Coussens, 1994). A direct relationship of these proteins or products of other ORFs and MDV oncogenicity has yet to be identified. However, some of these transcripts have been reported to prolong the proliferation and serum dependency of CEF, suggesting that they play an important role in the establishment and growth of tumour cells (Peng *et al.*, 1995).

MDV phosphoprotein 38 (pp38) complex

MDV-specific phosphoprotein pp38 was originally identified as an antigen expressed on tumour cells and lymphoblastoid cell lines (Cui *et al.*, 1990), suggesting a potential role in oncogenesis. It is a complex of two related phosphoproteins, pp24 and pp38, that are expressed during the cytolytic stages as well as in the feather-follicle epithelium (Cho *et al.*, 1998). Although pp38 can be detected in MDV-transformed lymphoblastoid cell lines and lymphomas, the expression is usually restricted to a small proportion of cells (Ross *et al.*, 1997), supporting the view that it is mostly associated with lytic stages rather than transformation. Furthermore, deletion of pp38 in the vvMDV strain Md5, reconstituted using overlapping cosmid technology (Reddy *et al.*, 2002), resulted in a virus that was severely restricted in early lytic replication *in vivo* in infected birds, even though *in vitro* replication was unaffected. In spite of the impairment in *in vivo* replication, the virus retained low levels of oncogenicity, demonstrating that pp38 is dispensable for tumour induction. Despite these observations, studies using antisense oligonucleotides directed against pp38 transcripts have indicated that pp38 does play a role in the proliferation of lymphoblastoid cells (Xie *et al.*, 1996).

MDV-encoded CXC chemokine, vIL-8

Another gene (vIL-8), which encodes a CXC chemokine, is located within the R_L region downstream of Meq (Liu *et al.*, 1999a; Parcells *et al.*, 2001). vIL-8 is a spliced gene product and is expressed as a late gene, although its expression is sometimes found in T-cell lines and tumours as well. vIL-8 has the highest homology with cellular IL-8 and GROa, although a highly conserved ELR sequence immediately preceding the CXC motif, considered to be important for neutrophil chemoattraction and angiogenesis, is missing in vIL-8 (see Chapter 10). Functional analysis has shown that vIL-8 is a secreted molecule and serves to attract chicken peripheral blood monocyte cells such as lymphocytes and macrophages and, to a lesser extent, heterophils. *In vivo*, vIL-8 seems to recruit target cells for MDV, leading to further expansion of infected cells and thereby increasing the likelihood of lymphoma development. This is consistent with data obtained using the vIL-8 knockout virus (Parcells *et al.*, 2001), where it was shown that this virus replicates well in tissue culture but with greatly reduced virus titres *in vivo* due to poor infection of lymphoid organs. Despite its reduced replicative function, the vIL-8 knockout virus is weakly oncogenic, and cell lines derived from tumours harboring latent MDV without vIL-8 have been isolated. While recruitment of B and T cells is likely to be an important function of vIL-8 in MDV replication, vIL-8 may contribute to oncogenesis in other ways. The chicken has several IL-8-related genes, including 9E3 and K-60 (Bedard *et al.*, 1987; Sick *et al.*, 2000). 9E3 was originally cloned as a gene that is overexpressed by v-*src* transformed cells, and was shown to be able to induce angiogenesis and proliferation of chicken embryo cells. By the same token, it is possible that vIL-8 may induce signals in T cells to maintain the activated state and enhance survival during viral replication. If it mimics cellular IL-8, vIL-8 may also activate several kinase cascades, including Src and the EGF-receptor. These kinases are known to activate the Jun/Fos pathway, and thus may act synergistically with Meq. Finally, vIL-8 could play a role in immunomodulation by interfering with the cellular IL-8 signal pathway, e.g. as a decoy or by formation of an inactive dimer.

MDV-encoded telomerase RNA

Recently, another gene encoding the RNA telomerase subunit (vTR), unique to MDV, has also been identified within the R_L region (Fragnet *et al.*, 2003). Telomerase is important for the maintenance of telomeres above a minimum-length threshold – a feature very important for actively proliferating transformed cells. The telomerase complex consists of two factors: a protein component TERT that has the features of reverse transcriptase, and the closely-associated RNA component TR, which acts as a template for TERT. Thus the identification of an MDV-encoded vTR could be significant in the induction and maintenance of transformation of lymphocytes by MDV. The sequences of vTR showed about 88 per cent homology to the chicken telomerase RNA (cTR). Functional analysis has shown that vTR can reconstitute telomerase activity in heterologous cells by interacting with the host cell telomerase protein. vTR is recognized only in serotype 1 MDV, strengthening the case for its association with the oncogenic properties of the virus. However, the presence of highly homologous vTR sequences (99.4 per cent) in both the oncogenic RB1B strain and the non-oncogenic CVI988 strain suggest that it may be important in oncogenicity but not as a major determinant. Nevertheless, further analysis of the role of vTR in viral replication and cell transformation could prove very interesting.

MDV ICP4 homologue-related transcripts

Among the immediate-early gene products of herpesviruses, ICP4 (RS1, MDV084) serves as a major transactivator that plays a crucial role in the regulation of transcription of many early and late genes. The ICP4 genomic region is within the R_S region and has very large potential ORF with proteins of sizes varying between 80 and 210 kDa (Xing *et al.*, 1999). Some of these proteins may have different localization within cells (Barrow *et al.*, 2003), although the significance is not clear. In spite of its role in activating several viral genes, a direct role for ICP4 in latency and transformation is yet to be demonstrated. Nevertheless, detection of high levels of ICP4-related transcripts in latently infected cells, as well as the inhibition of proliferating of lymphoblastoid cells by treatment with ICP4 antisense oligonucleotides, suggests some unidentified functions in latency and transformation.

The importance of ICP4 in MDV pathogenesis was demonstrated by a family of LAT antisense to ICP4 in lymphoblastoid cells and lymphomas (Cantello *et al.*, 1994; Li *et al.*, 1994; McKie *et al.*, 1995). These antisense transcripts comprise a large polyadenylated 10-kb L RNA and non-polyadenylated highly-spliced small antisense RNA (SAR/MSR) (Cantello *et al.*, 1994; Li *et al.*, 1994; McKie *et al.*, 1995) and three highly spliced 3′ co-terminal polyadenylated RNAs. SAR/MSR are abundantly expressed in transformed cells and lymphomas (Ross *et al.*, 1997). Although the mechanisms for generation of SAR/MSR are unclear, their non-polyadenylated features and predominant nuclear location suggest that they could be stable introns derived from a large RNA precursor (Cantello *et al.*, 1997). The pattern of their expression and inverse relationship with the sense transcripts suggests that they function as molecular switches for turning off MDV replication during latency. A mutant of RB1B virus, in which the 5′ end of the MSR at the LAT promoter region was disrupted, failed to induce tumours in experimentally infected or contact-exposed birds, suggesting a role for these transcripts in oncogenicity. However, firm conclusions on this will require further studies.

Summary

The quest to unravel the molecular mechanisms of MDV oncogenicity is at an important stage. The technologies for manipulating MDV genomes using BAC clones or reconstituted overlapping cosmids have revolutionized approaches for dissecting the functions of individual viral genes, or motifs within genes.

The application of rapid, targeted or random mutagenesis techniques should enable us rapidly to delineate the molecular mechanisms and determinants for MDV-induced oncogenesis.

References

Anderson, A. S., Parcells, M. S. and Morgan, R. W. (1998). *J. Virol.*, **72**, 2548–2553.

Askovic, S. and Baumann, R. (1997). *J. Virol.*, **71**, 6547–6554.

Bader, A. G., Hartl, M. and Bister, K. (2000). *Virology*, **270**, 98–110.

Barrow, A. D., Burgess, S. C., Baigent, S. J. *et al.* (2003). *J. Gen. Virol.*, **84**, 2635–2645.

Baxevanis, A. and Vinson, C. (1993). *Curr. Opin. Gen. Dev.*, **3**, 278–285.

Bedard, P. A., Alcorta, D., Simmons, D. L. *et al.* (1987). *Proc. Natl Acad. Sci. USA*, **84**, 6715–6719.

Bos, T. J., Monteclar, F., Mitsunobu, F. *et al.* (1990). *Genes & Development*, **4**, 1677–1687.

Brunovskis, P., Qian, Z., Li, D. *et al.* (1996). In *The 5th International Symposium on Marek's Disease, Kellogg Center, Michigan State University, East Lansing, Michigan, September 7–11*, pp. 265–270, AAAP, Kennett Square, Pennsylvania.

Call, K. M., Glaser, T., Ito, C. Y. *et al.* (1990). *Cell*, **60**, 509–520.

Calnek, B. W., Schat, K. A., Ross, L. J. and Chen, C. L. (1984). *Intl J. Cancer*, **33**, 399–406.

Cantello, J. L., Anderson, A. S. and Morgan, R. W. (1994). *J. Virol.*, **68**, 6280–6290.

Cantello, J. L., Parcells, M. S., Anderson, A. S. and Morgan, R. W. (1997). *J. Virol.*, **71**, 1353–1361.

Chang, K. S., Lee, S. I., Ohashi, K. *et al.* (2002). *J. Vet. Med. Sci.*, **64**, 413–417.

Chen, H., Lin, R. J., Schiltz, R. L. *et al.* (1997). *Cell*, **90**, 569–580.

Chen, H., Lin, R. J., Xie, W. *et al.* (1999). *Cell*, **98**, 675–686.

Cho, K. O., Park, N. Y., Endoh, D. *et al.* (1998). *J. Vet. Med. Sci.*, **60**, 843–847.

Cui, Z. Z., Yan, D. and Lee, L. F. (1990). *Virus Genes*, **3**, 309–322.

Damania, B. and Jung, J. U. (2001). *Adv. Cancer Res.*, **80**, 51–82.

Fragnet, L., Blasco, M. A., Klapper, W. and Rasschaert, D. (2003). *J. Virol.*, **77**, 5985–5996.

Fu, S., Bottoli, I., Goller, M. and Vogt, P. K. (1999). *Proc. Natl Acad. Sci. USA*, **96**, 5716–5721.

Glover, J. and Harrison, S. (1995). *Nature*, **373**, 257–261.

Hadman, M., Lin, W., Bush, L. and Bos, T. J. (1998). *Oncogene*, **16**, 655–660.

Hickabottom, M., Parker, G. A., Freemont, P. *et al.* (2002). *J. Biol. Chem.*, **277**, 47197–47204.

Himly, M., Foster, D. N., Bottoli, I. *et al.* (1998). *Virology*, **248**, 295–304.

Hong, Y. and Coussens, P. M. (1994). *J. Virol.*, **68**, 3593–3603.

Iwai, K., Mori, N., Oie, M. *et al.* (2001). *Virology*, **279**, 38–46.

Jones, D. and Kung, H. J. (1992). In *The 19th World's Poultry Congress, Amsterdam The Netherlands, September 19–24*, pp. 58–61, Ponsen and Looijen, Wageningen.

Jones, D., Lee, L., Liu, J. L. *et al.* (1992). *Proc. Natl Acad. Sci. USA*, **89**, 4042–4046.

Kim, H., You, S., Foster, L. K. *et al.* (2001a). *Oncogene*, **20**, 5118–5123.

Kim, H., You, S., Kim, I. J. *et al.* (2001b). *Oncogene*, **20**, 2671–2682.

Kung, H. J., Xia, L., Brunovskis, P. *et al.* (2001a). In *Marek's Disease* (ed. K. Hirai), pp. 245–260. Springer–Verlag, Berlin.

Kung, H. J., Xia, L., Brunovskis, P. *et al.* (2001b). *Curr. Topics Microbiol. Immunol.*, **255**, 245–260.

Lee, D. K., Kim, B. C., Brady, J. N. *et al.* (2002). *J. Biol. Chem.*, **277**, 33766–33775.

Lee, L. F., Wu, P., Sui, D. *et al.* (2000a). *Proc. Natl Acad. Sci. USA*, **97**, 6091–6096.

Lee, S. I., Takagi, M., Ohashi, K. *et al.* (2000b). *J. Vet. Med. Sci.*, **62**, 287–292.

Levy, A. M., Izumiya, Y., Brunovskis, P. *et al.* (2003). *J. Virol.*, **77**, 12841–12851.

Li, D. S., Pastorek, J., Zelnik, V. *et al.* (1994). *J. Gen. Virol.*, **75(7)**, 1713–1722.

Liang, M. H., Geisbert, T., Yao, Y. *et al.* (2002). *J. Virol.*, **76**, 4022–4033.

Liu, J. L. and Kung, H. J. (2000). *Virus Genes*, **21**, 51–64.

Liu, J. L., Lee, L. F., Ye, Y. *et al.* (1997). *J. Virol.*, **71**, 3188–3196.

Liu, J. L., Ye, Y., Lee, L. F. and Kung, H. J. (1998). *J. Virol.*, **72**, 388–395.

Liu, J. L., Lin, S. F., Xia, L. *et al.* (1999a). *Acta Virol.*, **43**, 94–101.

Liu, J. L., Ye, Y., Qian, Z. *et al.* (1999b). *J. Virol.*, **73**, 4208–4219.

Liu, J. L., Hebert, M. D., Ye, Y. *et al.* (2000). *J. Cell Sci.*, **113**, 1543–1552.

Louie, M. C., Yang, H. Q., Ma, A. H. *et al.* (2003). *Proc. Natl Acad. Sci. USA*, **100**, 2226–2230.

McKie, E. A., Ubukata, E., Hasegawa, S. *et al.* (1995). *J. Virol.*, **69**, 1310–1314.

Morgan, I. M., Havarstein, L. S., Wong, W. Y. *et al.* (1994). *Oncogene*, **9**, 2793–2797.

Mori, N., Fujii, M., Iwai, K. *et al.* (2000). *Blood*, **95**, 3915–3921.

Mori, N., Fujii, M., Hinz, M. *et al.* (2002). *Intl J. Cancer*, **99**, 378–385.

Nazerian, K. and Sharma, J. M. (1975). *J. Natl Cancer Inst.*, **54**, 277–279.

Neiman, P. E., Ruddell, A., Jasoni, C. *et al.* (2001). *Proc. Natl Acad. Sci. USA*, **98**, 6378–6383.

Parcells, M. S., Dienglewicz, R. L., Anderson, A. S. and Morgan, R. W. (1999). *J. Virol.*, **73**, 1362–1373.

Parcells, M. S., Lin, S. F., Dienglewicz, R. L. *et al.* (2001). *J. Virol.*, **75**, 5159–5173.

Peng, F., Bradley, G., Tanaka, A. *et al.* (1992). *J. Virol.*, **66**, 7389–7396.

Peng, Q., Zeng, M., Bhuiyan, Z. A. *et al.* (1995). *Virology*, **213**, 590–599.

Pessah, M., Marais, J., Prunier, C. *et al.* (2002). *J. Biol. Chem.*, **277**, 29094–29100.

Petropoulos, C. J. and Hughes, S. H. (1991). *J. Virol.*, **65**, 3728–3737.

Qian, Z., Brunovskis, P., Lee, L. *et al.* (1996). *J. Virol.*, **70**, 7161–7170.

Reddy, S. M., Lupiani, B., Gimeno, I. M. *et al.* (2002). *Proc. Natl Acad. Sci. USA*, **99**, 7054–7059.

Ross, N., O'Sullivan, G., Rothwell, C. *et al.* (1997). *J. Gen. Virol.*, **78**, 2191–2198.

Schat, K. A., Chen, C. L., Shek, W. R. and Calnek, B. W. (1982). *J. Natl Cancer Inst.*, **69**, 715–720.

Schat, K. A., Chen, C. L., Calnek, B. W. and Char, D. (1991). *J. Virol.*, **65**, 1408–1413.

Schuermann, M., Hunter, J., Hennig, G. and Muller, R. (1991). *Nucl. Acids Res.*, **19**, 739–746.

Sick, C., Schneider, K., Staeheli, P. and Weining, K. C. (2000). *Cytokine*, **12**, 181–186.

Stavnezer, E., Gerhard, D. S., Binari, R. C. and Balazs, I. (1981). *J. Virol.*, **39**, 920–934.

Stavnezer, E., Brodeur, D. and Brennan, L. A. (1989). *Mol. Cell Biol.*, **9**, 4038–4045.

Treier, M., Staszewski, L. M. and Bohmann, D. (1994). *Cell*, **78**, 787–798.

Tulman, E. R., Afonso, C. L., Lu, Z. *et al.* (2000). *J. Virol.*, **74**, 7980–7988.

Tusher, V. G., Tibshirani, R. and Chu, G. (2001). *Proc. Natl Acad. Sci. USA*, **98**, 5116–5121.

Wagner, S. and Green, M. R. (1993). *Science*, **262**, 395–399.

Witter, R. L. (1983). *Avian Dis.*, **27**, 113–132.

Witter, R. L. (1997). *Avian Dis.*, **41**, 149–163.

Xie, Q., Anderson, A. S. and Morgan, R. W. (1996). *J. Virol.*, **70**, 1125–1131.

Xing, Z., Xie, Q., Morgan, R. W. and Schat, K. A. (1999). *Acta Virol.*, **43**, 113–120.

Zhao, L. J. and Giam, C. Z. (1992). *Proc. Natl Acad. Sci. USA*, **89**, 7070–7074.

Marek's disease: a worldwide problem

5

CHRIS MORROW* and **FRANK FEHLER****
*Aviagen Ltd, Newbridge, Midlothian, Scotland
**Lohmann Animal Health GmbH & Co., Cuxhaven, Germany

Introduction

Despite the fact that the first description of a paralysing syndrome in four roosters, now recognized as classical Marek's disease (MD), by József Marek is close to its centenary, the consequences of infection with the virus responsible did not have an important impact on the poultry industry until the 1960s (see Witter and Schat, 2003). In the intervening years, poultry industries around the world grew dramatically. Intensification in poultry production was made possible by grain surpluses, developments in large-scale egg incubation techniques, use of artificial lighting to allow production throughout the year and control of *Salmonella pullorum* infection, as well as other technological advances that all combined to allow tremendous increases in poultry populations. The two major consequences of this expansion were the reduction in the genetic diversity of commercially produced poultry, and changes in their environment. The continuous availability of large populations of naïve hosts, which could be targeted with little effect on the survival of the virus, may have uncoupled the host–parasite relationships that were in existence until then. These changes could also have favoured the development of new virus strains with increased virulence.

During the first half of the twentieth century, the classical form of MD with paralysis of wings and legs was predominant. However, by the 1960s the lymphoproliferative form of the disease predominated and, since it had a much more rapid course, this was described as acute MD. Acute MD caused tumours in young layers (6–16 week of age) with up to 60 per cent mortality. Due to the short lifespan of the broiler (at that time 9 weeks), mortality from MD in meat-type birds was low. However, in the USA meat inspection condemnations from visceral and skin tumours became a problem. In flocks with extended rearing times, dramatic losses of up to 30 per cent were evident. This initial increase in virus virulence occurred before vaccination was introduced. By the late 1960s there were major advances in our understanding of the causes of MD and the definitive differentiation of this disease from lymphoid leukosis and similar conditions (see Chapter 2).

After introduction of the first MD vaccines, mainly the serotype 3 herpesvirus of turkeys (HVT) vaccine, field problems from MD appeared to be controlled. Losses decreased to acceptable levels, with tumour development being reduced or delayed in vaccinated flocks. However, the vaccine was not capable of inducing immune responses that protected against infection and shedding, although these were decreased in the vaccinated birds. The consequence of this was the presence of a continuous virus reservoir in these flocks. This continuously circulating virus population provided a base for the selection and adaptation of new MD virus (MDV) strains. These isolates obtained during the late 1970s were found to be able to break the protection induced by the first-generation HVT vaccines, especially those that were used as cell-free vaccines. In Europe, a serotype 1 vaccine CVI988 (Rispens *et al.*, 1972) was introduced and effectively controlled the field problems. It has since been adopted to control MD around the world (see Table 5.1).

During the second decade after the introduction of MD vaccines, an increasing number of vaccine breaks were reported and explained as an adaptation of the field viruses to selective pressures caused by widespread and intensive vaccination. In Europe, a bivalent vaccine consisting of both serotypes 1 and 3

Table 5.1 Year of introduction of CVI988 (Rispens) vaccine in different countries.

Country	Introduction	Comments
The Netherlands	1971	Other serotype 1 vaccines are not significantly used
Italy	1972	
Israel	1976	
Germany	1985	First registered
Canada	Early 1990s	
United Kingdom	1991	Was initially licensed for use in export hatcheries
Hungary	1993	
USA	1996	Other vaccines also in use
Korea	1996	
Australia	1997	CVI988 C/R6 strain was unsuccessfully tried in 1994
India	Not licensed	Serotype 2 (SB-1) & 3 (HVT) vaccines still in use
Peru	Not licensed	Serotype 2 (SB-1) & 3 (HVT) vaccines still in use
Chile	Not licensed	Serotype 2 (SB-1) & 3 (HVT) vaccines still in use
Cuba	Not licensed	Serotype 2 (SB-1) & 3 (HVT) vaccines still in use

was chosen to deal with the emerging disease problems, and the strategy of using bivalent vaccination successfully controlled the disease situation. In the USA, a comparable situation arose, but a different vaccination strategy using a bivalent vaccine comprising of serotype 2 (SB-1 strain) and HVT, was used to control the disease. However, in subsequent years MD losses continued to increase, necessitating introduction of the CVI988 vaccine for controlling MD in the mid-1990s. Comparison of virulence of MDV strains isolated from the USA during the last three decades has shown a remarkable continuum in the virulence characteristics, as judged by the ability to break immunity induced by different types of vaccines. Based on these properties, isolates have been classified into different pathotypes (virulent, vMDV; very virulent, vvMDV, and vv+MDV) with the vv+ viruses at the high end of the continuum (Witter, 1997). In Europe, on the other hand, no further increases in virulence have been reported since the last problem phase spanning the end of the 1980s and beginning of the 1990s. There are strains that could be considered vv+MDV, but these isolates continue to be well controlled under field conditions by use of CVI988 (Rispens) vaccine. Other manifestations of MD observed by field veterinarians include the condition referred to as *transient paralysis* in layers and *floppy chick syndrome* in broilers, both conditions being preventable by effective vaccination.

Diagnostic investigation of MD problems

For the field veterinarian diagnosis of MD can be difficult, and the tools to investigate outbreaks are time-consuming, of a specialist nature and expensive (see Chapter 12). The first stage in an investigation is confirmation that the problem is indeed MD. Until recently, the field veterinarian investigating MD problems has had histopathology as the first step in laboratory confirmation of field observations. Although virus isolation methods are available, they do not provide much more information because of the ubiquitous nature of the virus. Isolated viruses need to be characterized and further investigation needs experimental challenge studies in chickens (see Chapter 12), which are more difficult to carry out. The changing epidemiology of the disease has complicated diagnosis, with some veterinarians reluctant to consider MD where tumours were not seen until after 16 weeks (considered a diagnostic feature of avian leukosis virus (ALV)-induced lymphoid leukosis (LL)). MD outbreaks in broiler breeders are more likely to start with significant numbers of tumours appearing at the beginning of lay. This means that the lag time in controlling any problems can be six months, and hence any action taken cannot properly be judged effective because of the delay.

Case study 5.1 illustrates proventricular tumours in adult broiler breeders.

> **Case study 5.1** Proventricular tumours in adult broiler breeders
>
> Tumours starting at 22 weeks of age caused increased mortality in broiler breeders in Indonesia (1 per cent mortality per month). At *post mortem* the most common tumour was in the proventriculus, followed by the liver and spleen, with lung, kidney and other sites also being affected in some birds. No bursal or bone marrow tumours were seen, although a few ovarian tumours were observed. On histological examination, MD was confirmed.
>
> There was a history of change of MD vaccine supplier to a European manufacturer that did not sell MD vaccines within Europe. This decision was made on the basis of cost savings. The problem disappeared when the original vaccine was reinstated. In the end, no cost saving was achieved from switching to the cheaper vaccine.
>
> Late-onset tumour outbreaks, compared to classical MD outbreaks, are now very common, suggesting that MDV challenge does not occur until after transfer from the rearing farm (although sometimes outbreaks do occur without such a history).

The emergence of ALV subgroup J (ALV-J)-induced myeloid leukosis (ML) has caused confusion with the diagnosis of tumours in broiler breeder flocks. MD tumours have been diagnosed as ML, and this continues to be a problem in some areas. Host responses to some rapidly growing tumours include the infiltration of inflammatory cells that can be mistaken for cells of the myelocytic series; consequently, MD may not always be considered in the diagnosis. (Similarly, LL and ML can also be misdiagnosed as MD.) Clear descriptions of these problems are discussed by Barnes (1996), but are often ignored. Chronic inflammatory reactions may also be misdiagnosed as MD. In more remote parts of the world, there have been some major commercial problems caused by the misinterpretation of one tumour slide. Clearly there is a need for better quality assurance systems for avian histopathologists to allow correct diagnosis. Availability of standard reference materials and slide sets from experimental as well as field infections would help to improve their skills.

Another approach for preventing MD problems is the introduction of best practice at all stages in the production process. Such practices include: auditing the storage, reconstitution and administration of vaccines; implementation of standard operating procedures (SOPs) for the various methods; checking the compatibility of additives to MD vaccines (other vaccines, diluents and antibiotics); and use of colouring dyes when administering vaccines to confirm vaccine uptake. Some birds are now delivered from the hatcheries with a certificate that includes such details as the vaccine strain, name of the manufacturer and batch number of the vaccines used. Such practices achieve better quality control.

In many situations, the reason for an MD outbreak remains unclear. For every case study reported here, there have been more cases where the factors triggering MD outbreaks have not been identified. There have also been instances where the problems in a flock (or a house) disappear without any change in control measures.

Economic impact

Over the last 5 years, the MD situation has been quite stable in most areas of the world. Comprehensive reports on the worldwide MD situation are difficult to obtain (see Chapter 14 for the results of a recent global survey on the MD situation). There are different reasons for this:

- MD is not a notifiable disease.
- Low level losses after MD vaccination are generally accepted and treated as normal, since it is known that vaccination failures occur at low frequency.
- Occurrence of MD is often linked to financial claims between rearing companies and hatcheries or hatcheries and vaccine manufacturers. Often such cases are not made public.
- Since prevention of the disease requires optimal hygiene and management, besides several other measures, many MD cases are not reported in order to avoid damaging the reputation of the company concerned.

Furthermore, the number of unreported cases of MD is extremely high. This fact seriously undermines the evaluation of worldwide economic losses from MD. The economic impact of MD on the world poultry industry is thought to be in the range of US$1–2 billion annually, although this is a crude estimate and the real source of this value is not known. The Food and Agriculture Organization (FAO) estimates a total number of 45 billion broilers and 57 million tonnes of eggs (corresponding to about 900 billion eggs or about 5 billion laying birds) produced in the year 2002, with a total value of about US$100–200 billion. This means that the damage estimate given above corresponds to an MD-induced loss of about 1 per cent of the value. The figures are not verifiable, although they are readily quoted by researchers embarking on investigations into MDV infection. Reading the multi-annual animal disease status report for MD published by the Office International des Epizooties (OIE) clearly demonstrates that pathogenic strains of MDV are present in nearly every country with a developed poultry industry. A significant number of the total population, some 50 billion chickens produced yearly, is vaccinated against the disease, indicating the continual global threat of MD.

In summary, it can be said that there are still many problems concerning MD in the field. These differ from country to country. For instance, countries in southern Europe (such as Spain) often have severe problems with MD in meat-type birds, while others (such as France and Germany) currently struggle with increased MD mortality in layers. Some other countries have general problems in all sectors of intensive poultry management.

MD vaccination of broilers at a day old or *in ovo* may be carried out to increase performance of the broilers and improve carcass quality (reducing skin and visceral tumours) and yield, rather than to prevent overt clinical disease. This intervention strategy was pioneered in the USA (along with the practice of re-using litter), but is slowly spreading throughout the world. In the USA, of the 8550 million broilers inspected in 2002, the total condemnation rate was 0.8 per cent (0.4 per cent whole bird, 0.4 per cent parts), with the average live weight being 2.33 kg. The condemnations attributed to MDV generally account for 0.001 per cent in most regions, although the incidence of MD in the Delmarva region (0.011 per cent) is always higher than in other regions. Essentially all broilers in the USA are vaccinated against MDV, most with HVT and SB-1 bivalent vaccine administered at one-quarter to one-third of the recommended dose. Japan grows large broilers (2.8 kg average weight) and had a broiler condemnation rate of 0.13 per cent attributable to MD in 2000.

Causes of outbreaks

The list of possible causes is long, but some of the important ones include the following:

1. The main reason for the occurrence of increased mortality is the improper handling of the vaccines currently in use. MD vaccines are amongst the most delicate vaccines in veterinary use, and this is especially the case with the widely used cell-associated vaccines. They must be kept at −196°C during transport and until used. Even thawing the ampoules at temperatures of more than 28°C, or prolonged incubation before dilution, can cause damage to (or destruction of) the vaccine virus. In many developing countries (especially parts of Asia and Africa), improper handling frequently causes problems. In these areas, difficult transportation routes, hot climates and poorly educated vaccination personnel exacerbate the situation. It is not unusual for larger quantities of vaccine to be diluted than can be safely administered within a 2-hour period. In some cases, the vaccine from one supplier is diluted with diluents provided by another supplier, or the vaccine is mixed with additional vaccines (against other viruses) or with antibiotics. These combinations may not be compatible. Monitoring of improper handling is often difficult, since there are no practical field tests to confirm successful vaccination.

2. Hygiene is the second important weapon in the fight against MD. MDV is spread in dust particles containing MDV shed with dead epithelial cells from the feather follicles of infected birds. Sometimes young birds are introduced into houses where infected birds have been reared, before proper

cleaning and disinfection has been carried out. Lack of good hygiene and biosecurity measures within the houses is probably one of the main reasons for the development of MD problems in the USA. On many farms, disinfection of the accommodation between consecutive rearing periods is not commonly carried out. Vaccination against MDV does not lead to a sterilizing immunity – this means the animals are protected from clinical disease but not from infection, and shedding of MDV from the feather follicles can still occur. Pathogenic viruses are released with keratinized squamous epithelial cells even from vaccinated birds, and these viruses threaten unvaccinated chicks at the same location. The stability of such viruses in dry dust can extend for more than a year. These cell-free viruses are present only in the feather follicle cells, and it is thought that the keratin can protect the virus from destruction in the poultry house for long periods of time.

3. The preceding explanations make it clear how important good flock management is for controlling MD. Dramatic losses can occur, especially at locations where chickens of different ages are housed. The following scenario is typical. Older chickens on the production site have been vaccinated and show no clinical signs of MD, but they can shed field strains of virus. When young chicks (especially those less than 1 week of age) are introduced onto the same site, they are still susceptible to infection by MDV-infected dust because neonatal vaccination will not have had time to induce full protection against clinical MD. A competition begins between the vaccine virus and field virus within the host. Increased mortality may result and this may continue in subsequent generations introduced onto the site. Most European chicken integration operations use the 'all in–all out' approach, minimizing the reservoir of MDV field isolates since all animals within a flock are developing immunity to MDV at the same time. Together with proper hygiene arrangements before and after each generation is housed, the infection risk can be minimized. Nevertheless, some residual MD problems may still remain (see Case study 5.2).

Case study 5.2   Third placement shows increase in MD incidence

On large broiler parent flock complexes in Italy, chicks placed over 2 weeks have a history of increasing MD in younger flocks. This is despite revaccination with CVI988 (Rispens) vaccine on arrival or at 7 days of age. Could this be because of a build up of wild-type MDV increasing the early challenge for subsequent placements?

---

4. One poorly understood problem associated with MD outbreaks is the influence of other pathogens that cause immunosuppressive effects in the host (see Chapter 11). Examples of such pathogens are infectious bursal

disease virus (IBDV), avian leukosis virus (ALV), reoviruses, and the chicken infectious anaemia virus (CIAV; see Case study 5.3). The latter especially has become significant in many regions both within and outside Europe. During the last few years, many cases of MD have been investigated as part of an EU Framework 5 project on the development of new generation vaccines against MDV. In 11 of 15 cases from 8 countries from which proper material from acutely affected animals was available, CIAV could be identified as a co-infecting virus. The clinical signs accompanying this co-infection occurred predominantly in young layers before or directly at the point-of-lay. The MD-specific mortality occurred normally in these flocks until a certain

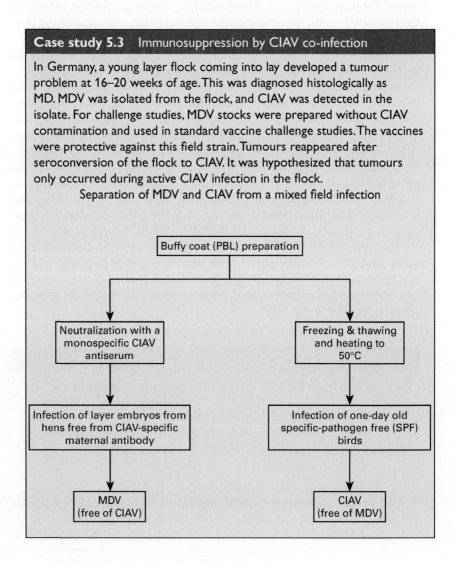

Case study 5.3   Immunosuppression by CIAV co-infection

In Germany, a young layer flock coming into lay developed a tumour problem at 16–20 weeks of age. This was diagnosed histologically as MD. MDV was isolated from the flock, and CIAV was detected in the isolate. For challenge studies, MDV stocks were prepared without CIAV contamination and used in standard vaccine challenge studies. The vaccines were protective against this field strain. Tumours reappeared after seroconversion of the flock to CIAV. It was hypothesized that tumours only occurred during active CIAV infection in the flock.

Separation of MDV and CIAV from a mixed field infection

Buffy coat (PBL) preparation

Neutralization with a monospecific CIAV antiserum

Freezing & thawing and heating to 50°C

Infection of layer embryos from hens free from CIAV-specific maternal antibody

Infection of one-day old specific-pathogen free (SPF) birds

MDV (free of CIAV)

CIAV (free of MDV)

stage. Thereafter, the morbidity and mortality increased abruptly. Transient paralysis in these flocks was rare, although dramatic weight losses were observed frequently. The frequency of tumours was relatively low, but classical symptoms such as proventriculus swelling (possibly from vagal nerve paralysis) were evident.

This novel phenomenon can be traced back to the immunosuppressive effects of CIAV. Usually it is considered that CIAV does not cause pathogenicity in older birds, only playing a role in young birds lacking maternal antibodies. Hence vaccination of layers was considered to be unnecessary. On the other hand, the immunosuppressive effects of field viruses are well known (Jeurissen *et al.*, 1992; Adair, 2000). Since MDV causes latent infection in unvaccinated as well as in vaccinated birds, and is controlled by cell-mediated immune surveillance, introduction of an immunosuppressive virus could disturb the crucial balance between MDV and the immune system, leading to an outbreak of clinical disease. This explanation of events needs to be investigated with other known immunosuppressive pathogens.

5. One further reason, often unappreciated, for increased MD mortality is the presence of stress. Many of the herpesviruses are known to become reactivated from latency by exposure to stressful situations. If this is also the case with MDV, the problems that induce stress need to be addressed (see Case study 5.4). It is common knowledge that stress induction readily occurs in chickens. Lack of food or drinking water and exposure to high temperature or mycotoxins are only a few examples of factors that induce stress. Moreover, it has been shown that stress can severely compromise immune responses in a chicken (El-Lethey *et al.*, 2003). Case studies 5.4 and 5.5 provide examples.

6. The evolution of new MDV variants must be considered the likely cause of MD outbreaks, as has been reported in the USA (Witter, 1997). Whether

---

Case study 5.4   Management stress increasing MD tumour incidence

A broiler breeder flock kept in the old sheds of a farm in Croatia was delayed coming into lay. At 26 weeks there was increased mortality due to MD tumours, and no eggs had been produced (under normal management conditions, the first eggs should be laid around 23 weeks of age). On examination of the lighting programme, taking into account the time of year and efficacy of light-proofing of the shed, it was concluded that decreasing day length was being sensed by the birds and was the cause of the increased incidence of MD tumours. Sister flocks in other countries had no tumour problem, indicating that the hatchery vaccination against MD was not the cause of the problem. Thus management should be aimed at minimizing any type of stress in birds.

these new variants have increased virulence or have just acquired the potential to evade the immune responses of the vaccinated birds in a flock is still being debated. The development of such variants could be the result of selective pressures caused by various practices in the poultry house environment. One important pressure is vaccination, especially if subprotective doses of vaccines are administered due either to improper usage or to overdilution of the vaccines, as is practised on some farms. A field infection occurring before establishment of protective immune responses can result in increased replication of the field virus in vaccinates, providing substantial genetic potential for adaptation to the selective pressure.

Case study 5.5   Interference in development of MD immunity

Birds were placed on a farm in the Czech Republic. The birds had been vaccinated in the hatchery with CVI988 (Rispens)/HVT, and were planned to be placed in four sheds. However, the last shed was not ready on the delivery day. Two sheds were used as usual, but the third shed had twice the number of birds planned (double stocking) without extra feeding equipment or water distribution for four weeks. The average first-week bodyweight in the double-stocked shed was markedly depressed (80 g compared with a 120 g target), but when the flock was divided after the four weeks, the chickens in sheds 3 and 4 were grown on to the body weight target. At 18 weeks the birds were transferred to laying farms. On the first laying farm, birds from sheds 1 and 2 performed normally. Birds derived from sheds 3 and 4 (all brooded in shed 3) developed MD, with 1 per cent of the birds dying each week from 23 weeks of age. Surplus birds from sheds 3 and 4 had also been transferred to another farm, and had no mortality or production problems (presumably no challenge). This seems to indicate that reasonable early management is required for immunity to develop during the first 4 weeks of age.

Genetic resistance of the host is another factor that affects MDV evolution, especially when the resistance is determined by only one gene. With MDV, many host genes could be involved in genetic resistance, but the mechanisms are not clear. If each gene functions by a different mechanism, then each also produces its own selective pressure on the virus (discussed in detail in Chapter 9). Knowledge of genetic resistance to MD has been exploited in some commercial breeding programmes (McKay, 1998).

The influence of the factors cited here is increased by high population densities, which favour the rapid replication and horizontal spread of MDV. Mistakes in management practices assist in creating an environment that helps the pathogen to evolve. In summary, the situation in a given MD outbreak is usually not simple. A retrospective analysis by outsider observers is not normally possible, since many factors have to be taken into account to identify the reasons for the outbreak and to take appropriate remedial measures.

In many developing countries, such as parts of Asia and Africa, there are severe problems with education in hygiene and proper procedures for vaccine handling. Together with the hot climates often present in these countries, these factors can lead to massive MD losses. Due to high temperatures and dry conditions, the risks of infection are high. In addition, vaccination practices often are not properly performed, use of disinfection measures is rare, and multi-age farm models are still in use. Moreover, in developed countries effective hygiene practices are not applied. The MD outbreaks in the late 1990s in the USA occurred on farms where routine disinfection between consecutive production runs was not practised. In most developed countries, the reasons for various MD outbreaks are not clear. Problems may occur in a certain flock, or house, and disappear without any history of changes in husbandry practices. The description of vv+MDV pathotypes by Witter (1997) in many countries created some concern. Since then, practitioners have looked more carefully for MD cases and have naturally found them. The current situation is highly complicated. Many veterinarians have agreed to administer several MD vaccinations. Some administer two vaccinations within the first day of life in order to reduce the risk of vaccination failure, while others prefer to revaccinate at 7 or 14 days of age to obtain a booster effect. Normally the first vaccination is done with a bivalent HVT/Rispens vaccine, while for re-vaccination a monovalent Rispens vaccine is used. Currently there is no scientific evidence to confirm that this practice is beneficial, but some field veterinarians are convinced that the current low incidence of MD in many countries is mainly attributed to the double vaccination.

MDV infection as a primary cause of immunosuppression is also an important consideration in the field (see Case study 5.6). Detailed descriptions of the immunosuppressive effects of MDV and other viruses are given in Chapter 11.

Case study 5.6 MDV as a suspected cause of primary immunosuppression

In Australia, some flocks (vaccinated against coccidiosis) were reported to have developed *Eimeria acervulina* coccidiosis around 22 weeks of age, and often later went on to suffer large losses from MD tumours. Whether this is indicative of a primary immunosuppressive effect of MDV or an underlying immunosuppression that triggers both the MD and the break in coccidial immunity is not clear. It is unlikely to be due to CIAV infection since these birds have been effectively vaccinated against CIAV at 11 weeks of age.

MD in species other than chickens

Although MD is primarily a problem in chickens, other avian species occasionally succumb to the disease (see Chapter 7). Reports of MD in quail (see, for example, Pennycott *et al.*, 2003) have described a picture very similar to that seen in many outbreaks involving chickens (infiltration of liver, spleen, proventriculus and small intestine but not peripheral nerves). There was concurrent candidiasis and megabacteriosis. The challenge in this case appears to come from vaccinated chickens. MD in turkeys is not currently a large problem and vaccination is not used except in some special production systems, such as the *Label Rouge* systems in France, which use turkey genotypes that are not used by conventional producers. HVT vaccine is not protective in turkeys, and CVI988 vaccine is most effective in preventing the disease. Sporadic MD is sometimes reported in pheasants and related species.

Problems associated with contamination of MD vaccines

Egg-drop syndrome-76, REV (Bagust *et al.*, 1979), ALV and CIAV have all been either suspected or proven to be contaminants of MD vaccines. The most severe clinical outbreaks of disease associated with some of these viruses (e.g. CIAV and REV) have been reported in the field when contamination of MD vaccines has occurred. The unique requirement of MD vaccines to be administered either to embryos or chicks immediately after hatching makes the diseases caused by other viruses a particularly serious problem; the age resistance to the pathogenic effects of these infections is not fully developed in these chicks at this time. Contamination of MD vaccines has been a major problem ever since the industry started to use them, and this needs to be addressed by the vaccine manufacturers and regulatory authorities. Amyloid arthropathy associated with *Enterococcus faecalis* resulting from the use of contaminated MD vaccines has been reported in brown leghorn chickens (Landman *et al.*, 1998).

Summary

The CVI988 (Rispens) vaccine was certainly the saviour of poultry industries worldwide during the 1990s. Questions as to whether all the CVI988 vaccine strains are equally effective will preoccupy researchers over the next decade. There have also been some concerns about reports of encephalitis induced by CVI988 vaccines in susceptible stocks (von Bülow, 1977), but this is not considered to be a field problem (Maas *et al.*, 1982). Over 10 million doses of CVI988 vaccine each year are used in broiler breeders in the UK, and there have been no reports of any encephalitis associated with this vaccine during the 10 years of its use. Likewise, it can safely be used in layer stocks. SB-1 (and other serotype 2 vaccines) has been shown to potentiate tumours caused by ALV (Bacon *et al.*, 1989) and REV infections in susceptible genotypes, but authorities have never discouraged the use of serotype 2 vaccines (although SB-1 is no longer available in Brazil).

Currently, around the world, there appear to be no major MD problems that are not being controlled with existing technologies. The question worrying poultry veterinarians is, will MDV continue to increase in virulence and become a problem again? Should there be fundamental changes to the organization of the industry that could protect the useful life of the CVI988 (Rispens) vaccine? An on-farm assessment of whether vaccination has been properly performed is needed. Techniques based on real-time quantitative PCR assays provide a good deal of promise for the future (see Chapter 12).

Field veterinarians still have many questions they would like the research community to answer. Are the MDV strains that infect broiler breeders and layers (birds with a long lifespan) the same as the strains that infect broilers? Where do new MDV strains evolve? Do strains of increasing virulence evolve in a locality, or do they spread between different countries? Although preliminary attempts have been made to understand why the virulence of field isolates has increased over the last 40 years, there is a greater need for understanding why this has happened. Will there be a need for new vaccines to deal with ever-increasing virulence in field strains of MDV? Are there changes to management systems that could prolong the usefulness of our current vaccines? Do we need vaccines that can prevent MDV infection by inducing sterile immunity? Finally, is poultry production, as we know it, sustainable with the co-existence of the virus and the host?

References

Adair, B. M. (2000). *Dev. Comp. Immunol.*, **24**, 247–255.

Bacon L. D., Witter, R. L. and Fadly, A. M. (1989). *J. Virol.*, **63**, 504–512.

Bagust, T. J., Grimes, T. M. and Dennett, D. P. (1979). *Aust. Vet. J.*, **55**, 153–157.

Barnes, H. J. (1996). In *Avian Histopathology* (ed. C. Riddell), 2nd edn, p. 12. American Association of Avian Pathologists, Pennsylvania.

El-Lethey, H., Huber-Eicher, B. and Jungi, T. W. (2003). *Vet. Immunol. Immunopathol.*, **95**, 91–101.

Jeurissen, S. H., Wagenaar, F., Pol, J. M. *et al.* (1992). *J. Virol.*, **66**, 7383–7388.

Landman, W. J., vd Bogaard, A. E., Doornenbal, P. *et al.* (1998). *Amyloid*, **5**, 266–278.

Maas, H. J. L., Borm, F. and van de Kieft, G. (1982). *World Poultry Sci. J.*, **38**, 163–175.

McKay, J. C. (1998). *Avian Pathol.*, **27**, S74–S77.

Pennycott, T. W., Duncan, G. and Venugopal, K. (2003). *Vet. Rec.*, **153**, 293–297.

Rispens, B. H., van Vloten, H., Mastenbroek, N. *et al.* (1972). *Avian Dis.*, **16**, 108–125.

von Bülow, V. (1977). *Avian Pathol.*, **6**, 395–403.

Witter, R. L. (1997). *Avian Dis.*, **41**, 149–163.

Witter, R. L. and Schat, K. A. (2003). In *Diseases of Poultry* (ed. Y. M. Saif), 11th edn, pp. 407–465. Iowa State Press, Iowa.

Marek's disease virus: biology and life cycle

6

SUSAN J. BAIGENT and FRED DAVISON

Institute for Animal Health, Compton Laboratory, Newbury, Berkshire, UK

Introduction

Like the other herpesviruses, the virion of MDV consists of an envelope comprising a complex, loose, lipid sac incorporating the viral glycoproteins, which surrounds an amorphous tegument (see Plate 1). Within the tegument, an icosahedral capsid encloses a linear double-stranded DNA core. Although the genome structure of MDV indicates that it is an α-herpesvirus like herpes simplex and varicella zoster virus (see Chapter 3), biological properties indicate MDV is more akin to the γ-herpesvirus group, which includes Epstein-Barr and Kaposi's sarcoma herpesvirus. These herpesviruses replicate lytically in lymphocytes, epithelial and fibroblastic cells, and persist in lymphoblastoid cells.

MDV has a complex life cycle (Plate 2) and uses two means of replication, productive and non-productive, to exist and propagate. The method of reproduction changes according to a defined pattern depending on changes in virus-cell interactions at different stages of the disease, and in different tissues. Productive (lytic) interactions involve active invasion and take-over of the host cell, resulting in the production of infectious progeny virions. For many herpesviruses, lytic infection is fully productive and results in release of cell-free virus independent of the infected cell type. However, some herpesviruses, including MDV, can also establish a semi-productive (abortive) infection in certain cell types, resulting in production of cell-associated progeny virus. MDV has a rapid replicative cycle of 18–20 hours. Attachment of virus to host cell receptors is likely to involve glycoproteins B, C and D (gB, gC and gD) and initiates fusion to the target cell, then penetration. Uncoating of the virion, by cellular enzymes, releases the viral DNA, which circularizes and enters the nucleus. Messenger RNA (mRNA), synthesized in the nucleus, is transported into the cytoplasm for translation. Viral gene expression occurs in a coordinately regulated, sequentially ordered fashion: proteins expressed from immediate early (IE) genes, such as infected cell proteins (ICP) ICP4 and ICP10, regulate expression of early (E) and late (L) genes. Approximately 50 per cent of the viral genome is expressed, and about 50 viral polypeptides are associated with lytic infection. DNA replication occurs between the early and late stages. DNA is cleaved into unit length molecules, and proteins necessary for capsid formation are transported to the nucleus, where encapsidation of the genome occurs (Figure 6.1a). The virus particles gain a tegument and, by 18 hours, assembled nucleocapsids are budding through the inner nuclear membrane, gaining an envelope (Figure 6.1b). Virions mature in the Golgi complex (Figure 6.1c), and are released from the cell by exocytosis from Golgi vesicles. Release of progeny virus is accompanied by changes in cell metabolism, extensive cytopathology and death of the target cells.

Non-productive interactions represent persistent infection, in which the viral genome is present but gene expression is limited, there is no structural or regulatory gene translation, no replication, no release of progeny virions and no cell death. Reactivation of the virus is rare and, usually, the infectious virus can be re-isolated only after cultivation *in vitro*. For the virus, this is an economical means of perpetuation and transmission, affording stability and protection from destruction by the host's immune system. There are two forms of non-productive infection; latency and transformation (see later).

Calnek and colleagues, working at Cornell University, postulated that there are four phases of MD pathogenesis in susceptible birds: an early cytolytic phase (2–7 days post-infection, or dpi), a latent phase (7–10 dpi onwards), a late cytolytic and immunosuppressive phase (18 dpi onwards) and a proliferative phase (28 dpi onwards). They formulated a hypothesis (reviewed by Calnek, 1986, 2001) to explain these sequential changes, and this 'Cornell Model' is the generally accepted hypothesis and is well supported by experimental evidence.

The outcome of MDV infection depends on a number of virus-determined and host-associated factors:

1. *Virus serotype and pathotype*. Serotype 2 MDV and serotype 3 (turkey herpesvirus, HVT) are naturally occurring, infectious viruses, but are non-pathogenic or only weakly pathogenic and non-oncogenic in chickens. Conversely, serotype 1 MDV are highly infectious, and can cause cytolytic infection of lymphocytes, with visceral lymphomas and neural lesions. Four pathotypes of serotype 1 MDV have been proposed by Witter (1997). This system is based on work at the Avian Disease and Oncology Laboratory, East Lansing, USA, using strains of MDV isolated from vaccine breaks in the USA (see Chapters 5 and 14); it has not yet been extended to other parts of the world where vaccines breaks have also been reported. Witter (1997) suggested that MDV isolates from the earlier half of the twentieth century probably were moderate pathotypes causing predominantly classical MD (see Chapter 7). With intensification of poultry production, acute MD caused by virulent MDV (vMDV) came to predominate. Compared with

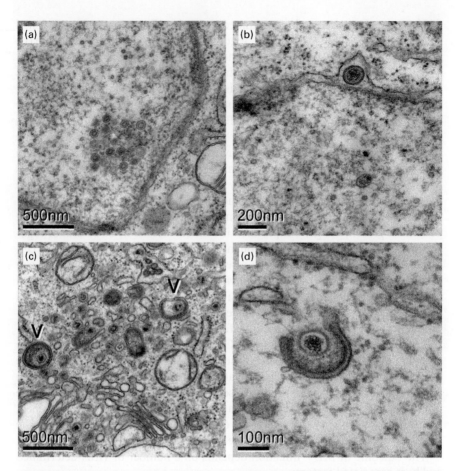

Figure 6.1 Electron micrographs showing different stages in the envelopment of MDV virions in a chicken embryo fibroblast that had been infected with MDV using BAC20. This BAC clone was constructed from vv+MDV (584 strain) that had been attenuated by 80 passages *in vitro* (Schumacher *et al.*, 2000). (a) Assembled MDV nucleocapsids can be seen in the nucleus of an infected fibroblast cell. (b) MDV particles can be seen budding through the nuclear membrane. (c) MDV virions can be seen in the cytoplasm near the Golgi region. Some virions (V) appear to be wrapped or being wrapped in a transGolgi vesicle. (d) A close-up of a mature MDV virion being wrapped in a double membrane. (Electron micrographs reproduced courtesy of Daniel Schumacher, Cornell Veterinary College.)

vMDV, very virulent (vvMDV) and very virulent plus (vv+MDV) have excessive pathogenicity in vaccinated chickens. vvMDV are isolates against which HVT vaccine provides poor protection, and vv+MDV isolates produce significantly higher levels of disease in bivalent-vaccinated chickens (Witter, 1997). Increasing virulence is associated with central nervous system manifestations, increased visceral lymphoma frequency, increased mortality, early mortality with bursal and thymic atrophy, and increased frequency of ocular lesions (Witter, 1983).

2. *Host genotype.* Although all chicken strains can be infected by MDV, they differ in their susceptibility to development of MD lesions, clinical signs and mortality, dependent on their genetic constitution. Two types of genetic resistance have been described: that associated with the major histo-compatibility complex (MHC), and that associated with genes outwith the MHC (see Chapter 9).

3. *Maternal antibodies.* The presence of maternally derived antibodies in young chicks can significantly protect them against early challenge with MDV.

4. *Vaccination. In ovo* or neonatal vaccination of chicks with HVT, serotype2/HVT bivalent vaccine, or CVI988 (Rispens) serotype 1 vaccine will protect against vMDV, vvMDV or vv+MDV respectively (see Chapter 13).

5. *Age at infection.* Neonatal chicks are highly susceptible to MD, but this susceptibility decreases with the development of a functional immune system, and a degree of 'age resistance' is observed. Susceptibility is also influenced by the stage of sexual development.

6. *Stress and immunosuppression.* Young chicks are particularly susceptible to husbandry-associated stresses such as transport, feed restriction, beak-trimming, heat, cold and dehydration. Additionally, concurrent infection by immunodepressive pathogens, such as chicken infectious anaemia virus, reovirus and infectious bursal disease virus, significantly enhances susceptibility to MD (see Chapter11), as does antibiotic treatment.

Natural infection with MDV

Marek's disease virus spreads by indirect contact and, in environments contaminated by infected birds, can remain infectious for several months. The source of infectious cell-free virus is the feather-follicle epithelium (FFE), the only site where fully productive infection and release of cell-free MDV occurs (see later). Virus is shed with the debris of dead stratified epithelial cells and also by moulted feathers with infected cells attached (Carrozza *et al.*, 1973). This MDV-infected poultry dander and dust can be inhaled by other chickens. Early studies showed that dust from MDV-infected poultry houses could infect clean experimental birds after 14 days, about the time that MDV antigens were detected in the skin of the infected birds. Infection could be transmitted either by whole dust or by virus isolated from the dust by high-speed centrifugation (Carrozza *et al.*, 1973), suggesting that MDV may exist in two different forms in feather dust. It was proposed that cell-free virus particles associated with skin debris are highly infectious but labile. Keratin-wrapped particles are less infectious but more stable in the environment, due to the protection afforded by the cell debris and other components of poultry dust.

Little is known about events following infection by this natural route, because of problems with delivering a known, precisely timed dose of MDV in its 'native' form in dust. Thus, most studies have been, and are currently, performed using parenteral administration of cell-associated MDV obtained from cell culture or peripheral blood leukocytes from MDV-infected chickens. There have been various attempts to mimic natural infection using suspensions of MDV-infected dust, but the approach lacks precision in delivering a quantifiable dose and timing. To avoid this, Davidson and Borenshtain (2003) used 'liquid' extracts of MDV-infected feather tips as a source of infectious MDV. Dripping the infectious extracts into the beaks of experimental chicks to initiate mucosal infection induced viraemia and tumours. Recently a method has been developed for insufflating MDV-infective dust into the lower trachea to cause MD (B. Baaten, C. Butter and T. F. Davison, IAH Compton, unpublished results).

The lung is almost certainly the portal for virus entry, although the actual site(s) of uptake along the airways and the cellular mechanisms involved in virus uptake have yet to be identified. However, the avian lung has a very different structure from its mammalian counterpart, being a more inflexible and open system connected to nine thin-walled avascular air sacs that act like giant bellows moving air through the lungs. Also the avian lung lacks alveoli, with gas exchange taking place in capillaries that form a mantle around the tertiary bronchi (parabronchi). Due to the narrowness of these capillaries there are few airway-resident macrophages, and it is unclear whether it is lung-resident macrophages or external macrophages that transport particles across the pulmonary epithelium. Macrophages can enter the blood circulation and also drain from the lung via deep lymphatic vessels to transport particulate matter to the lymphoid tissues.

In poultry dust, the MDV load is similar in different size fractions, indicating that virus-infected particles of dust can reach deep into the lungs (G. J. Underwood and T. F. Davison, unpublished observations). In addition, there may be different mechanisms for uptake of keratin-wrapped MDV and cell-free MDV. Which of these forms of dust-associated MDV is predominantly responsible for initiating infection remains to be determined. Free MDV could cross the pulmonary epithelium following binding to a receptor, while keratin-wrapped MDV may cross in the same manner as other particles following phagocytosis by macrophages. Studies are necessary to elucidate the mechanisms involved.

The subsequent pathogenesis of Marek's disease will be now be described based on the 'Cornell Model', although, as will become clear, there is recent evidence that this model may require further elaboration.

Early cytolytic phase

Some MDV, carried by macrophages from the lungs into the bloodstream, will enter the secondary lymphoid tissues. Expression of large amounts of the MDV immediate-early protein pp38 and presence of the MDV genome can be demonstrated in most secondary lymphoid tissues (the spleen, gut-associated lymphoid tissue, caecal tonsil, Harderian gland and conjunctiva-associated lymphoid tissue) from 2–7 dpi, peaking at 4 dpi, and is accompanied by lymphocytolysis, with necrosis and infiltration of inflammatory cells (see Chapter 7). It is likely that these tissues all become infected synchronously by the uptake of MDV-carrier cells from the circulation, the spleen and caecal tonsil being the preferential sites for uptake of migrating cells and for presentation of antigens. In the spleen the open-ended capillaries are surrounded by ellipsoid-associated reticular cells (EARCs), which can phagocytose virus. MDV antigens are readily detected in splenic EARCs (Jeurissen *et al.*, 1989a, 1989b) and splenic macrophages (Barrow *et al.*, 2003) during early infection. Recent studies (Barrow *et al.*, 2003) show that macrophages are not only phagocytic carriers of MDV, but that some MDV strains (e.g. hypervirulent C12/130) can also cytolytically infect and replicate in these cells. Splenic macrophages that expressed the MDV antigens ICP4, pp38 and gB had high inherent death rates indicative of cytolytic infection. Although it was not possible to observe virus particles, nuclear localization of ICP4 was consistent with infection rather than phagocytosis. It is quite likely that EARCs also become infected – in fact, MDV infection of EARCs and macrophages could well be a prerequisite for transmission of the virus from these carrier cells to lymphocytes, the major target cell.

In the spleen, B lymphocytes directly surround EARCs (Jeurissen *et al.*, 1989a, 1989b) and this could be the reason for them being the primary targets of acute cytolytic infection. Splenic B cells are important but not absolutely essential for MDV replication and amplifying the virus, and so have a crucial effect on disease outcome: high levels of cytolytic infection in B cells correlate with subsequent lymphoma development. However, removal of the spleen prior to infection delays the spread of cytolytic infection to other lymphoid tissues by 2–3 days, but neither influences the appearance of latently-infected blood lymphocytes nor abolishes visceral lymphoma development (Schat, 1981). This implies that the spleen is important, although not essential, for early pathogenesis but not for the development of sequential events.

Pathological responses to infection in the lymphoid organs are described in detail in Chapter 7. Infection and pathology in the primary lymphoid organs, bursa and thymus lag behind the spleen by 1 day, and the virus genome load is lower than in the spleen (G. J. Underwood and T. F. Davison, unpublished observations). It is likely that the source of infection of these tissues is cytolytically infected B cells entering the circulation from the spleen and caecal tonsil. It is unclear whether infection of the thymus, bursa and other lymphoid aggregates is important for virus amplification and disease pathogenesis, or whether their infection is a dead-end. However, infection can result in severe atrophy of the bursa and thymus (see Chapter 7), leading to immunosuppression, and with some vv+ pathotypes of MDV cytolytic infection can be devastating.

In each of the lymphoid tissues, the vast majority (~90 per cent) of cytolytically infected (MDV-antigen-positive) cells are B lymphocytes, with ~3 per cent being CD4$^+$ and CD8$^+$ TCR$\alpha\beta^+$ T lymphocytes increasing to ~6 per cent by 7 dpi (Shek *et al.*, 1983; Calnek *et al.*, 1984a; Baigent *et al.*, 1998). Although these viral antigens are first detected 3–4 dpi, the MDV genome can be detected in B and T lymphocytes as early as 2 dpi, with virus load being significantly higher in B cells.

It is generally accepted that resting T cells are fairly refractory to infection but that cytolytic infection of B cells induces activation of T cells, rendering these cells susceptible to infection (Calnek *et al.*, 1984b). It is conceivable that activation induces expression of an MDV receptor on T cells. The phenotype of the activated T cells is likely to be influenced by type of T cells present at the site of infection, as demonstrated by a local lesion model (Calnek *et al.*, 1989) in which MDV-infected chick kidney cells, inoculated into the wing-web of allogeneic birds, induced lymphocyte infiltration. Early lesions contained both B and T lymphocytes while, from 7 days onwards, the majority were CD4$^+$ and CD8$^+$ T cells. Close interaction between B and T cells, as part of the immune response, would facilitate MDV spread to T cells. However, it is possible that T lymphocytes can also be directly infected by macrophages since, in cultures of *in vitro*-infected spleen cells from embryonally bursectomized chicks, the majority of MDV-antigen-positive cells are T cells (Calnek *et al.*, 1984b). The actual percentage of lymphocytes that become cytolytically infected by MDV in each lymphoid organ is very low (less than 2 per cent in MD-susceptible and ~0.2 per cent in MD-resistant chickens, following infection with a vMDV), perhaps reflecting the cell-associated nature of the virus (Baigent *et al.*, 1998; Baigent and Davison, 1999).

Cytolytic infection of B and T cells is semi-productive, which may reflect an inability of these cells to make certain viral structural components, or degradation of mature virus during passage from cell membranes. No cell-free virus is produced, only non-enveloped intranuclear particles, so it is not known exactly how virus spreads from cell-to-cell. However, the highly conserved MDV envelope glycoproteins gH, gL, gE, gI and gM are indispensable for intercellular spread in cultured avian cells. gH and gL form a complex, as do gE and gI. gM putatively complexes with the homologue of HSV UL49.5 protein. Interaction of gH and gL, which probably occurs in the cytosol or ER lumen, is necessary for both gH and gL subcellular translocation and for cell surface expression. Expression of the gH/gL hetero-oligomer on the cell surface is required for viral spread to uninfected cells (Wu *et al.*, 2001). Mutant studies using an infectious bacterial artificial chromosome (BAC) clone of MDV-1 showed deleting gE or gI (Schumacher *et al.*, 2001), gM or UL49.5 (Tischer *et al.*, 2002) resulted in virus progeny that were unable to spread from cell-to-cell, reflected by the absence of virus plaques and the detection of only single infected cells following transfection. Of the MDV-1 major tegument proteins, only VP22, encoded by the UL49 gene, is essential for cell-to-cell spread (Dorange *et al.*, 2002), a function which may depend upon its ability to bind microtubules.

Latency

At 6–7 dpi, expression of MDV antigens in the lymphoid tissue is lost, marking the switch from cytolytic to latent infection. During MDV latency, the viral genome persists in the host cell although no viral or tumour antigens are expressed, and there is no production of infectious virus except after reactivation. MDV latency is difficult to study, as it cannot be established *in vitro*. *In vivo* samples can be contaminated with transformed cells, and it is difficult to distinguish latent infection from transformation, since both represent non-productive infections. Latently infected T cells may be those that eventually become transformed: the relationship between the two states is not fully understood. In many instances the MDV–host interaction during latency has been studied using MD lymphomas and cell lines, but it is perhaps better to study latency using MDV antigen-negative splenocytes taken 10–12 dpi, before transformation begins. There are three recognized stages to latency – establishment, maintenance and reactivation – but the virus–host relationship is dynamic and transition between these states can occur. The latent and lytic cycles diverge prior to the expression of IE genes and before DNA replication.

Several lines of evidence implicate the involvement of extrinsic, host-determined factors in the switch from cytolysis to latency. First, latency in T and B cells is broken (i.e. they enter semi-productive cytolytic infection) 12–30 hours after explantation *in vitro*, when they are freed from host influences. Secondly, latent infection is first observed 6–7 dpi, when the initial immune response is evident. Finally, re-emergence of cytolytic infection in the lymphoid organs and epithelia can be induced by immunosuppression. The view that host immune responses are crucial for establishment and maintenance of latency is supported by studies on immunocompetence (Buscaglia *et al.*, 1988). Chemically induced immunosuppression prior to infection with MDV results in a prolonged, widespread early cytolytic infection, while immunosuppression after latency has developed results in reactivation of cytolytic infection in the lymphoid tissues. Establishment of latency is perturbed by thymectomy but not by bursectomy, implying that the cell-mediated rather than the humoral immune response is crucial.

The influence of the immune system on the course of MD pathogenesis is described in Chapter 10. However, it is relevant to point out here that a latency-maintaining factor, which was present in conditioned medium from activated spleen cells and could maintain latency in MDV-infected spleen cells *in vitro*, has been described (Buscaglia and Calnek, 1988). More recent investigations have identified host cytokines (IL-6, IL-18, IFNγ) and soluble mediators (nitric oxide) that may contribute to the establishment and maintenance of latency. Nitric oxide (NO) may play an important role in the control of MDV replication. Replication of MDV in CEF cultures was inhibited by the induction of NO using recombinant chicken IFNγ and lipopolysaccharide; CEF or splenocytes derived from MD-resistant chickens produced NO earlier and at higher levels than did those from susceptible chickens (Xing and Schat, 2000). *In vivo*, the enzyme inducible NO synthase (iNOS), which synthesizes NO, was up-regulated between 6 and 15 dpi in MDV-infected birds, and treatment of these birds with an inhibitor of iNOS resulted in increased virus load.

Following *in vitro* reactivation of MDV from virus antigen-negative lymphocytes, the majority (~90 per cent) of latently infected lymphocytes were identified as T cells, with only ~3 per cent being B cells (Shek *et al.*, 1983; Calnek *et al.*, 1984a) and the predominant latently infected cell type is considered to be the CD4$^+$ TCR$\alpha\beta^+$ lymphocyte subset. According to the Cornell model of MD pathogenesis (Calnek, 1986), the switch to latent infection occurs 6–7 dpi in T and B lymphocytes that were previously cytolytically infected. However, alternative views deserve consideration. A subset of CD4$^+$ cells could become latently infected as early as 3 dpi, perhaps directly infected by macrophages, without requiring activation by infected B cells (G. J. Underwood, IAH, unpublished observations). This would be consistent with the local lesion model, described earlier, in which transformable T cells were present as early as 4 dpi (Calnek *et al.*, 1989). Cytolytic infection of B cells, and subsequently activated T cells, results in death of most of these cells so that, by 6–7 dpi, the MDV-infected cells which predominate are those CD4$^+$ cells that were latently infected from the outset. This would suggest that the cytolytic phase is not essential for subsequent MD pathogenesis. Consistent with this, embryonal bursectomy (which should remove all B cells) prior to MDV infection abolished cytolytic infection, while latent infection and lymphoma formation still occurred (Schat *et al.*, 1981). Clearly, further studies are required to determine the origin of the latently infected T cells.

The status of the viral genome in MDV latently infected and transformed cells is described later in the section on transformation, but transcription in latently infected cells is discussed here. During latency, transcription is limited to latency-associated transcripts (LATs), which are likely to be important in the balance between latent and lytic infections. MDV LATs include two MDV small RNAs (MSRs) and a 10-kb RNA that are detected in MDV-infected chicken fibroblasts only from 6 days (the onset of latency). Furthermore, the LATs are relatively abundant in MDV-transformed cells and are reduced in number upon reactivation of MDV (Cantello *et al.*, 1994, 1997). The LATs are a complex family of spliced RNAs, predominantly localizing to the nucleus. The MSRs contain two potential ICP4 recognition sequences and a latency promoter binding factor recognition sequence toward their 5′ ends. MDV LATs map antisense to the ICP4 gene, and are considered to interfere with translation of the major immediate early regulatory protein ICP4, thereby suppressing lytic infection (Cantello *et al.*, 1994, 1997). Use of 'quantitative reverse-transcription real-time PCR' demonstrated high levels of LATs and *Meq* transcripts in CD4$^+$ T splenocytes, and lower levels in CD8$^+$ T splenocytes and B splenocytes, taken from 5 dpi onwards from MDV-infected birds: with the onset of latency, the levels of these transcripts reach a plateau in CD8$^+$ and

B cells, while in CD4$^+$ cells they continue to increase (G. J. Underwood and T. F. Davison, unpublished observations), consistent with the fact that CD4$^+$ cells are the important latently infected subset *in vivo*. *Meq* plays an important role in maintaining latency, since it blocks apoptosis of latently infected CD4$^+$ T cells and transactivates latent gene expression (Parcells *et al.*, 2003). Upon reactivation of lytic infection, splice variants of *Meq* predominate and these lack the transactivation domains.

In resistant chicken genotypes, latent infection persists at a low level in the spleen and blood lymphocytes without further effect. However, in susceptible or suppressed birds, or those infected with a vvMDV pathotype, a second serious pathological cycle begins 2–3 weeks after primary infection. Latently infected peripheral blood lymphocytes disseminate the virus around the body to organs as diverse as the skin, viscera and nerves.

Late cytolytic phase

The inflammatory changes in the lymphoid organs, a direct response to cytolytic infection of B cells, resolve from 7–14 dpi. However, in MD-susceptible chickens, a second wave of semi-productive infection and cytolysis from 14–21 dpi has been reported (Calnek, 1986). This is coincident with permanent immunosuppression associated with lymphoma development. This late cytolytic infection affects the thymus, bursa and some epithelial tissues, including FFE (see below), kidney, adrenal gland and proventriculus (see Chapter 7). Necrosis of lymphocytes and epithelial cells is accompanied by pronounced inflammation, infiltration of mononuclear cells and heterophils and (for the bursa and thymus) severe atrophy. It is suggested that latently infected cells carry the virus to these tissues, where it becomes reactivated due to a secondary wave of immunosuppression (Calnek, 1986).

Fully productive infection in the feather-follicle epithelium

Virus is probably carried to the skin by latently infected peripheral blood lymphocytes, and can be detected 10–12 dpi. In the skin, productive infection is reactivated in FFE to produce gross lesions as lymphocytes aggregate around the infected follicles (see Chapter 7). From about 13 dpi, virus replication is fully productive, resulting in the release of infectious, enveloped, cell-free virus by passive cell break-up, accompanied by cell death. Enveloped particles can be observed in the cytoplasm as well as in the nucleus, and these might represent a more stable state than the nuclear particles seen in abortively infected cells. The FFE could be a privileged site for production and release of infectious cell-free virus (see earlier) due to the fact that protection is afforded by envelopment in cytoplasmic inclusion bodies, and gradual de-keratinization of cells reduces cytoplasmic lysosomal activity, protecting the virus from degradation. MDV glycoprotein D (gD) is likely to be important for production of cell-free virus, although it is not essential for virus replication in cell culture or in chickens (Niikura *et al.*, 1999). *In vivo*, gD is expressed only in the FFE, suggesting its expression is differentially regulated and associated with production of cell-free virus. In resistant and susceptible chickens alike, high levels of the MDV genome (S. Baigent, unpublished observations) and MDV antigens persist in the FFE, and cell-free virus is shed throughout the life of the bird.

Transformation

It is generally considered that neoplastic transformation of latently infected lymphocytes to lymphoblastoid tumour cells is the ultimate consequence of interaction of MDV with the host cell (see Chapter 8). The spleen is likely to be the predominant site for initial proliferation of transformed cells, but cannot be the only source of transformable target cells, as splenectomized birds still develop neoplastic MD lesions (Schat, 1981). Three weeks post infection the splenic T-dependent areas become hyperplastic, and following this diffusely distributed T cells, presumed to be precursors of neoplastically transformed cells, are seen throughout the spleen (Ichijo *et al.*, 1981; Baigent and Davison, 1999). As detailed in Chapter 8, cells expressing high levels of CD30, a host-encoded extracellular antigen expressed by MD tumours and cell lines, are first detected in the spleen and blood of both resistant and susceptible chickens at the end of the cytolytic phase (Baigent *et al.*, 1998; Burgess and Davison, 2002). CD30 is expressed at very low levels by small minorities of uninfected leukocytes (Burgess and Davison, 2002) and is not expressed on resting T cells, so the CD30$^+$ splenic T cells are likely to be MDV-infected precursors of MD tumour cells.

From 3–4 weeks post-infection, non-productively infected lymphocytes progressively migrate into the visceral organs and peripheral nerves, where, under the influence of as yet undetermined factors, they proliferate to form lymphomas (see Chapter 7 for pathology detail). Lymphocytes continually circulate from the blood to the lymphatics via lymphoid and non-lymphoid tissue, and migration through endothelial cell monolayers requires the production of matrix-degrading enzymes – an ability that increases in activated T cells (Masuyama *et al.*, 1992), such as those latently infected with MDV. It is not known whether the latently infected cells become transformed after reaching the visceral organs, where a locally produced factor might initiate transformation, or whether transformation has already occurred in the spleen or at some other site. However, transformation is likely to be a multi-step process characterized by the accumulation of multiple genetic events (see later).

About 75 per cent of cells prepared from visceral lymphomas are T lymphocytes, while only about 15 per cent are B lymphocytes (Rouse *et al.*, 1973; Payne and Rennie, 1976; Baigent, 1995). However, non-producer lymphoblastoid cell lines established from MD lymphomas (see Chapter 7) are 100 per cent CD4$^+$, TCRαβ$^+$ T cells, indicating that these cells represent the neoplastic component of lymphomas. Other cells in lymphomas *in vivo* (see Chapter 8) are non-infected inflammatory and immune cells, which might be reactive against transformed cells or against viral antigens (Payne and Rennie, 1976). Infiltration of TCRαβ$^+$ T lymphocytes into the visceral organs, initially forming small focal lesions around the blood vessels, occurs in both MD-resistant and MD-susceptible White Leghorn chickens from 15 days after intra-abdominal MDV injection (Baigent, 1995; Burgess *et al.*, 2001). MDV antigens cannot be detected in these organs at this time, so the infiltrating lymphocytes are unlikely to be part of an immune response to viral antigens. In susceptible birds there is extensive proliferation of CD4$^+$ TCRαβ$^+$ cells, which often invade and totally replace the normal tissue, forming mature lymphomas by 50 dpi. However, in resistant birds CD8$^+$ cells predominate and the lesions usually regress from 30 dpi, associated with apoptosis (Burgess *et al.*, 2001), suggesting that proliferation of MDV-transformed CD4$^+$ cells is kept in check by these CD8$^+$ cells, which perhaps have cytotoxic anti-tumour activity (Baigent, 1995). Although MD cell lines have provided important information on virus–cell interactions and the composition and development of lymphomas, it is conceivable that changes can occur during passage in tissue culture. Therefore, the study of freshly explanted lymphoma cells provides a better option. Burgess and Davison (2002) showed that all MD lymphomas contain sub-populations of cells that are low expressors of CD30, and sub-populations that are high expressors. The CD30^{hi} cells showed characteristics of neoplastically transformed cells (as detailed in Chapter 8) – namely protection from cell death despite hyperproliferation, and inability to support MDV productive infection – and they had the phenotype CD4$^+$, TCRαβ$^+$, MHCI^{hi}, MHCII^{hi}, IL-2Rα$^+$, CD28$^{lo/-}$, pp38$^-$, gB$^-$. Additionally, CD30$^+$ lymphoma cells had greater MDV loads than did CD30$^-$ lymphoma cells, which might be latently infected, but not transformed (Burgess and Davison, 1999, 2002). CD4$^+$, CD30$^+$ lymphoma cells express *Meq* and the small LAT MSR, consistent with the notion that these are the transformed component of MD lymphomas (Ross *et al.*, 1997).

Why are only CD4$^+$ T cells transformed under normal conditions of MD pathogenesis? Towards the end of cytolytic infection, both B and T cells are infected and the switch to latent infection is likely to occur simultaneously in B and T cells. Perhaps latently infected B cells cannot provide the appropriate environment for transformation. Integration of viral DNA into the host cell genome (see later) requires cell proliferation, a requirement easily met by the population of activated, latently infected T cells. There is evidence for the ability of MDV to transform other lymphocyte types under abnormal circumstances. Depletion of CD4$^+$ cells prevented tumour formation in three of four

MDV-infected birds, but tumours of non-T cell origin developed in the fourth bird (Morimura, 1993). This suggests that while the normal targets for transformation are CD4$^+$ T cells, another cell type may be transformed in depleted birds. That the phenotype of transformable target cells is affected by factors influencing the types of activated T cells at the site of infection was indicated by the local lesion model (Calnek *et al.*, 1989). Lymphoblastoid cell lines derived from the local lesions were CD3$^+$ TCRαβ$^+$ T cells; however, only 11 of these cell lines were CD4$^+$ while 23 were CD8$^+$ and 19 were CD4$^-$CD8$^-$ (Schat *et al.*, 1991).

Payne and Rennie (1976) suggested that the neoplastic component of tumours is monoclonal – that is, a lymphoma arises from a single transformed cell that proliferates at that site, and not from the progeny of several transformed cells. In agreement with this, different lymphomas within the same bird have the same MDV genome copy number and integration site pattern, whereas this varies between different birds (Delecluse *et al.*, 1993). However, recent analysis of TCR-vβ gene expression suggests that MD lymphomas have polyclonal origins (Burgess and Davison, 2002). Thus, it is conceivable that a balance between the host's immune responses and MDV-transformed cells determines the number of integrated MDV genomes. As detailed in Chapter 8, cells that are latently infected or transformed by MDV show persistence of the viral genome. In low-producer cell lines a minor population of latently infected cells contain free extrachromosomal, circular (episomal) MDV DNA and support the lytic cycle (Kaschka-Dierich *et al.*, 1979; Delecluse and Hammerschmidt, 1993). However, non-producer cell lines, which do not support the MDV lytic cycle, have only integrated copies of the MDV genome at multiple sites (Kaschka-Dierich *et al.*, 1979; Delecluse and Hammerschmidt, 1993), as do primary MD lymphoma cells (Delecluse *et al.*, 1993). The relationship between latently infected cells and transformed cells is not totally clear; however, the co-existence of episomal and integrated copies of the MDV genome in some cell lines shows that MDV infection is a precursor of lymphoma development. It is conceivable that in latently infected cells the genome is episomal, and that integration is a prerequisite for transformation. While sites of integration may be random, eventual transformation appears to be associated with integration near the telomeres of the host's minichromosomes and larger chromosomes (Delecluse and Hammerschmidt, 1993; Moore *et al.*, 1993). The number of copies of the MDV genome per transformed cell is greater than in latently infected cells, having been determined at 10–20 copies per transformed cell (Ross *et al.*, 1981) and 2–12 copies (Delecluse and Hammerschmidt, 1993). Integrated MDV DNA is treated as cellular DNA, under cellular control, and is replicated once per cell cycle in the early S-phase, just before cellular DNA, so all progeny cells carry the integrated genome. Integration of MDV might act to inhibit cytolytic gene transcription and enhance the expression of genes associated with latency (LATs) and transformation (Meq). The short latency of MDV and the polyclonal nature of lymphomas is consistent with MDV having its own potent oncogenes (see Chapter 4).

The latent/transformed state might be influenced by methylation of the MDV genome. MDV DNA in cell lines is considerably methylated at 5′ CpG 3′ dinucleotides, and prevention of methylation by treatment with 5-azacytidine results in enhanced MDV antigen expression and viral replication (Fynan *et al.*, 1993), an altered pattern of transcription and a change in the nucleosomal structure of the genome (Hayashi *et al.*, 1995). Transcription of mRNAs in MDV lymphoma cells and in non-producer cell lines is limited to regions in or near the repeat sequences of the MDV genome (Sugaya *et al.*, 1990; Ohashi *et al.*, 1994). Less than 20 per cent of the MDV genome is transcribed, with approximately 65 per cent of the resulting RNA being transported to ribosomes. This prevents viral replication and maintains the state of cellular transformation. The few transcripts and their gene products may be important for initiation or maintenance of transformation. To the right of the MDV origin of replication are genes associated with early lytic infection (including pp38/pp24), while to the left are genes associated with latent/transforming infection (including the 1.8-kb RNA family, and Meq). In MD cell lines, viral transcripts include the 1.8 kb-mRNA, which might be involved in viral oncogenicity. The long inverted repeats of the serotype 1 MDV genome contain closely located transcriptional promoters for pp38/pp24 and 1.8 kb-mRNA, which initiate transcription in opposite directions and are regulated simultaneously by binding of a viral or a cellular factor induced by infection (Shigekane *et al.*, 1999). The relative abundance of these 'opposing' transcripts might contribute to initiation of the transformed state. It is conceivable that transformation occurs after a set duration, or that increasing Meq transcription in latently infected cells results in transformation when a threshold for Meq expression is reached, as discussed in the 'Model for MD lymphomagenesis' in Chapter 8.

Conclusions

The well established model for the pathogenesis of MD (the Cornell Model) describes the disease in terms of four phases: the early cytolytic phase, the latent phase, the late cytolytic and immunosuppressive phase, and the proliferative phase. Our current understanding of these sequential stages of the MDV life cycle (Plate 2) has been described in this chapter. While data from recent studies in essence support the Cornell Model, they also provide evidence that the real situation may be more complex. Contemporary studies should soon shed more light on many aspects of the disease. These include the early stages of infection via the respiratory route; the importance of macrophages in disseminating MDV around the body; the role of early T-cell infection in the subsequent formation of lymphomas; and the function of Meq in MDV latency and transformation. However, there still will be gaps to fill in our knowledge of the biology and life cycle of this fascinating herpesvirus, and, with developments in genome and protein analysis, answers should be forthcoming in the not too distant future.

Acknowledgements

We are very grateful to Daniel Schumacher, Cornell University College of Veterinary Medicine, Ithaca, USA, for providing the electron micrographs of MDV and giving us permission to use them. We also wish to thank Steve Archibald, Mick Gill and Paul Monaghan for help with the graphics and micrograph interpretation.

References

Baigent, S. J. (1995). PhD thesis, University of Bristol, UK.

Baigent, S. J. and Davison, T. F. (1999). *Avian Pathol.*, **28**, 287–300.

Baigent, S. J., Ross, L. J. and Davison, T. F. (1998). *J. Gen. Virol.*, **79**, 2795–2802.

Barrow, A. D., Burgess, S. C., Baigent, S. J. *et al.* (2003). *J. Gen. Virol.*, **84**, 2635–2645.

Burgess, S. C. and Davison, T. F. (1999). *J. Virol. Methods*, **82**, 27–37.

Burgess, S. C. and Davison, T. F. (2002). *J. Virol.*, **76**, 7276–7292.

Burgess, S. C., Basaran, B. H. and Davison, T. F. (2001). *Vet. Pathol.*, **38**, 129–142.

Buscaglia, C. and Calnek, B. W. (1988). *J. Gen. Virol.*, **69**, 2809–2818.

Buscaglia, C., Calnek, B. W. and Schat, K. A. (1988). *J. Gen. Virol.*, **69**, 1067–1077.

Calnek, B. W. (1986). *CRC Crit. Rev. Microbiol.*, **12**, 293–319.

Calnek, B. W. (2001). *Curr. Topics Microbiol. Immunol.*, **255**, 25–56.

Calnek, B. W., Schat, K. A., Ross, L. J. N. and Chen, C.-L. H. (1984a). *Intl J. Cancer*, **33**, 399–406.

Calnek, B. W., Schat, K. A., Ross, L. J. N. *et al.* (1984b). *Intl J. Cancer*, **33**, 389–398.

Calnek, B. W., Lucio, B., Schat, K. A. and Lillehoj, H. S. (1989). *Avian Dis.*, **33**, 291–302.

Cantello, J. L., Anderson, A. S. and Morgan, R. W. (1994). *J. Virol.*, **68**, 6280–6290.

Cantello, J. L., Parcells, M. S., Anderson, A. S. and Morgan, R. W. (1997). *J. Virol.*, **71**, 1353–1361.

Carrozza, J. H., Fredrickson, T. N., Prince, R. P. and Luginbuhl, R. E. (1973). *Avian Dis.*, **17**, 767–781.

Davidson, I. and Borenshtain, R. (2003). *FEMS Immunol. Med. Microbiol.*, **38**, 199–203.

Delecluse, H. J. and Hammerschmidt, W. (1993). *J. Virol.*, **67**, 82–92.

Delecluse, H. J., Schuller, S. and Hammerschmidt, W. (1993). *EMBO J.*, **12**, 3277–3286.

Dorange, F., Tischer, B. K., Vautherot, J. F. and Osterrieder, N. (2002). *J. Virol.*, **76**, 1959–1970.

Fynan, E. F., Ewert, D. L. and Block, T. M. (1993). *J. Gen. Virol.*, **74**, 2163–2170.

Hayashi, M., Io, K., Furuichi, T. *et al.* (1995). *J. Vet. Med. Sci.*, **57**, 157–160.

Ichijo, K., Isogai, H., Okada, K. and Fujimoto, Y. (1981). *Zbl. Vet. Med. B.*, **28**, 177–189.

Jeurissen, S. H. M., Janse, E. M., Kok, G. L. and De Boer, G. F. (1989a). *Vet. Immunol. Immunopathol.*, **22**, 123–133.

Jeurissen, S. H. M., Scholten, R., Hilgers, L. A. T. *et al.* (1989b). *Avian Dis.*, **33**, 657–663.

Kaschka-Dierich, C., Nazerian, K. and Thomssen, R. (1979). *J. Gen Virol.*, **44**, 271–280.

Masuyama, J.-I., Berman, J. S., Cruikshank, W. W. *et al.* (1992). *J. Immunol.*, **148**, 1367–1374.

Moore, F. R., Schat, K. A., Hutchison, N. *et al.* (1993). *Intl J. Cancer*, **54**, 685–692.

Morimura, T. (1993). *Jpn J. Vet. Res.*, **41,** 34–35.

Niikura, M., Witter, R. L., Jang, H. K. *et al.* (1999). *Acta Virol.*, **43,** 159–163.

Ohashi, K., O'Connell, P. H. and Schat, K. A. (1994). *Virology*, **199,** 275–283.

Parcells, M. S., Arumugaswami, V., Prigge, J. T. *et al.* (2003). *Poultry Sci.*, **82,** 893–898.

Payne, L. N. and Rennie, M. (1976). *Avian Pathol.*, **5,** 147–154.

Ross, N. L., DeLorbe, W., Varmus, H. E. *et al.* (1981). *J. Gen. Virol.*, **57,** 285–296.

Ross, N. L., O'Sullivan, G., Rothwell, C. *et al.* (1997). *J. Gen. Virol.*, **78,** 2191–2198.

Rouse, B. T., Wells, R. J. H. and Warner, N. L. (1973). *J. Immunol.*, **110,** 534–539.

Schat, K. A. (1981). *Avian Pathol.*, **10,** 171–182.

Schat, K. A., Lucio, B. and Carlisle, J. C. (1981). *Avian Dis.*, **25,** 996–1004.

Schat, K. A., Chen, C. L., Calnek, B. W. and Char, D. (1991). *J. Virol.*, **65,** 1408–1413.

Schumacher, D., Tischer, B. K., Reddy, S. M. and Osterrieder, N. (2001). *J. Virol.*, **75,** 11307–11318.

Shek, W. R., Calnek, B. W., Schat, K. A. and Chen, C.-L. H. (1983). *J. Natl Cancer Inst.*, **70,** 485–491.

Shigekane, H., Kawaguchi, Y., Shirakata, M. *et al.* (1999). *Arch. Virol.*, **144,** 1893–1907.

Sugaya, K., Bradley, G., Nonoyama, M. and Tanaka, A. (1990). *J. Virol.*, **64,** 5773–5782.

Tischer, B. K., Schumacher, D., Messerle, M. *et al.* (2002). *J. Gen. Virol.*, **83,** 997–1003.

Witter, R. L. (1983). *Avian Dis.*, **27,** 113–132.

Witter, R. L. (1997). *Avian Dis.*, **41,** 149–163.

Wu, P., Reed, W. M. and Lee, L. F. (2001). *Arch. Virol.*, **146,** 983–992.

Xing, Z. and Schat, K. A. (2000). *J. Virol.*, **74,** 3605–3612.

Pathological responses to infection

7

LAURENCE N. PAYNE

Formerly Institute for Animal Health, Compton Laboratory, Newbury, Berkshire, UK

Introduction

Infection of susceptible chickens by virulent Marek's disease virus (MDV) leads to pathological changes that comprise the condition termed Marek's disease (MD). These changes characteristically involve lymphoid tissues, especially from lymphoma formation, but nervous tissues and a variety of other tissues are often also affected. The pathology of MD is complex, and the pattern of the lesions seen has varied from the time that József Marek first described the disease in 1907. The changing pathology, it now seems, is mainly a result of changes in the virulence and tropism of the virus, although the mutations responsible have yet to be defined. The disease described by Marek was considered a peripheral polyneuritis, and this neural form (causing 'fowl paralysis') was later termed 'classical' MD by Biggs (see Chapter 2). In 1926, Pappenheimer and his colleagues reported that lymphomas, particularly in the ovary, were sometimes associated with the neural form. These workers considered the disease to be neoplastic, and termed it 'neurolymphomatosis gallinarum'. In the 1950s and 1960s the lymphomatous expression in MD became much more prevalent, and this virulent form was termed 'acute leukosis' or 'acute MD'. Since then, increasingly severe lymphomatous disease has been observed.

Before the experimental transmission of MD in the early 1960s, and the discovery of the herpesvirus responsible for it in 1967 (see Chapter 2), descriptions of the pathology of MD were based on naturally occurring field cases. The ability to induce the disease experimentally provided a source of material of known provenance, and allowed studies to be made on the sequence of pathological events occurring after infection. The ability to track the sites of virus replication (for example by viral immunofluorescence) and to characterize the nature of the cellular responses in MD (for example by use of monoclonal antibodies against lymphoid cell and accessory cell subsets) has allowed significant progress to be made in understanding the pathogenesis of the MD lesions. Currently, progress is also beginning to be made in understanding the molecular pathogenesis of the disease (see Chapter 4).

In this chapter, literature citations are provided mainly for the more recent advances in knowledge. For older references, the reader is referred to review articles on MD pathology and pathogenesis, including those by Payne *et al.* (1976), Payne (1985), Schat (1987), Calnek (2000) and Witter and Schat (2003).

Virus–cell relationships in Marek's disease

Knowledge of the types of virus–cell interactions that occur in MD is crucial for understanding the pathogenesis and pathology of the disease (see Chapter 6). These interactions are common to a number of members of the family *Herpesviridae*, and include:

1. *Fully productive infection*, characterized by production of fully-infectious virions, accompanied by cell death, and occurring in MD only in infection of feather-follicle epithelium.
2. *Semi-productive infection*, in which viral antigens and naked nuclear virions are formed and in which infectivity is cell-associated. This leads to cell death and is exemplified in MD in infection of lymphoid tissues and, to a lesser extent, of parenchymatous tissues.
3. *Non-productive latent infection*, in which viral genome persists in lymphoid cells that express no viral antigens.
4. *Non-productive neoplastic infection*, in which viral genome persists in lymphoid cells, with limited antigenic expression, and results in the immortalization of cells. These are typified in MD in lymphoma cells, transplantable lymphomas and lymphoma-derived lymphoid cell lines, and are discussed in Chapter 8.

Pathogenesis and pathology of Marek's disease

The following account is based largely on the sequential histopathological changes that occur when newly-hatched, MD maternal antibody-free, genetically susceptible chicks are infected with a virulent strain of MDV, which leads to the definitive lesions. This scheme is illustrated in Figure 7.1, and the changes are discussed as far as possible in chronological order.

Primary infection

Under natural conditions, MDV infection is usually acquired by inhalation of infective feather debris in the poultry house. Viral antigen is detectable in the

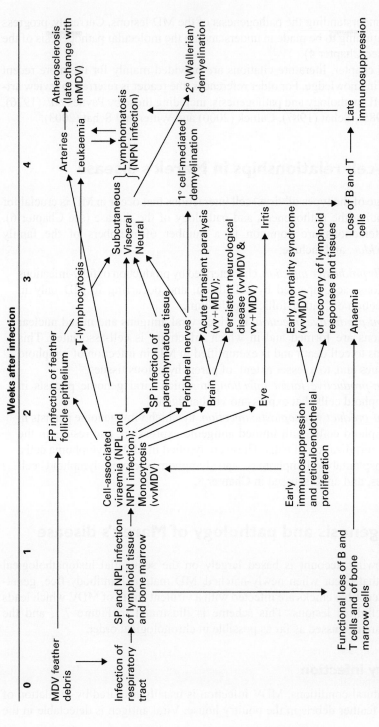

Figure 7.1 Pathogenesis of Marek's disease based largely on infection with vMDV of young, MD maternal antibody-free, genetically susceptible chicks. Also included are lesions mainly associated with strains of MDV of other pathotypes (mMDV, vvMDV, vv+MDV). Types of infection: FP, fully productive; SP, semi-productive; NPL, non-productive latent; NPN, non-productive neoplastic.

lung at 24 hours post-infection. Although recent evidence using quantitative RT-PCR assays suggest that the lung lymphoid tissues support MDV replication (Bas Baaten, IAH, personal communication), no lung pathology is associated with this phase of the disease.

Acute cytolytic infection of lymphoid tissue

An acute cytolytic infection of lymphoid tissue, notably in the bursa, thymus and spleen, appears at 3 days after infection, reaches a peak at 5–7 days, and usually resolves by 2 weeks (Payne and Rennie, 1973) (Figure 7.2). B cells are the primary target cells in these organs (Shek *et al.*, 1983). $CD4^+$ and $CD8^+$ T cells then become activated and infected, and a few activated, infected $CD4^+$ T cells transform into neoplastic lymphoma cells. This sequence of events is discussed in detail in Chapter 6.

The acute cytolytic infection is of the semi-productive type, with abundant viral antigen and cell-associated virus present. There is marked cytolysis of B and T lymphocytes (Baigent and Davison, 1999) and reticulum cells, and numerous intranuclear herpesvirus inclusion bodies may be seen. The virus infection provokes an acute to subacute inflammatory response in these organs, with granulocytic invasion and an increase in reticulum cells and macrophages. In the bursa

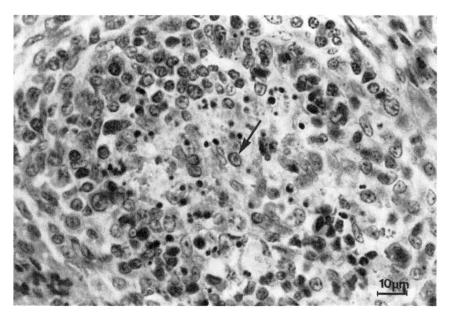

Figure 7.2   Acute cytolytic infection by vMDV in a lymphoid follicle of the bursa of Fabricius, 7 days after infection of a 1-day-old chick. An intranuclear inclusion body (arrow) is present. (From Payne, 1985, with kind permission.)

and thymus these changes are accompanied by severe regression of bursal lymphoid follicles and thymic cortex, resulting in a weight loss of these organs (see Chapter 11). Thymocytes also undergo massive apoptosis (Morimura *et al.*, 1996). In the spleen, regressive lymphoid changes do not occur and the inflammatory changes result in a weight increase. These changes usually resolve, and by 2 weeks, the architecture of these lymphoid organs is almost restored.

Baigent and Davison (1999) made a detailed study of lymphocyte subset responses in the spleen of susceptible (line 7_2) chicks infected with vMDV. They described, at 4–5 days post-infection, a 'lymphoid lesion' structure in which patches of MDV-infected and uninfected B cells became surrounded by $CD8^+$ $TCR\alpha\beta^+$ (possibly cytolytic lymphocytes reacting against infected B cells), which in turn were surrounded by $CD4^+TCR\alpha\beta^+$ cells (perhaps immune helper cells). These aggregates were then surrounded by $TCR\gamma\delta^+$ cells (perhaps involved in down-regulating the immune response prior to virus latency). Latent infection of $CD4^+$ cells is then believed to lead to their neoplastic transformation.

Certain very virulent (vv) strains of MDV, when inoculated experimentally, can cause severe acute cytolytic disease and bursal and thymic atrophy, and result in death of the bird in this phase (see Chapter 11). This is termed the 'early mortality syndrome' (Witter *et al.*, 1980; Barrow and Venugopal, 1999).

Viraemia and haematology

Infection by MDV is followed by an early and persistent cell-associated viraemia. The virus is present in buffy coat cells (apparently mainly in lymphocytes), but is not free in the plasma. Such infected cells are believed to transport virus to other tissues. The infection is of the semi-productive and latent types.

The bone marrow becomes infected with MDV by 5 days after infection, and in chicks without MDV maternal antibody this can lead to marrow aplasia and severe anaemia. An extravascular haemolytic anaemia resulting from erythrophagocytosis of red blood cells by hyperplastic reticuloendothelial cells was observed in the liver and spleen of chicks infected with the AC-1 and RB-1B strains of MDV (Gilka and Spencer, 1995).

During the acute cytolytic stage of infection with a virulent (v) MDV (HPRS-16), Payne and Rennie (1976a) observed in the blood an increase in absolute numbers of B cells, T cells, total lymphocytes and heterophils, and a decrease in numbers of monocytes and eosinophils. During the later lymphoproliferative phase, T cells and total lymphocytes increased, resulting in a leukaemia in some birds, and B cells, monocytes, heterophils and basophils decreased. Apoptosis of $CD4^+$ T cells and down-regulation of CD8 T cell molecules was observed in the blood, associated with MDV-induced immunosuppression (Morimura *et al.*, 1995).

In a recent study with the highly virulent C12/130 strain of MDV, Barrow *et al.* (2003) observed a marked peripheral blood monocytosis (about a 10-fold increase) peaking at 8 days post-infection. $CD4^+$ and $CD8^+$ cells decreased in the blood between 4–7 days post-infection, during the acute cytolytic stage, and then increased above normal. The increases in these cell types coincided with severe perivascular cuffing in the brain and signs of acute transient paralysis (see *Central nervous system* below).

Infection of feather-follicle epithelium

Feather-follicle epithelium (FFE) is the one location in the chicken in which a fully productive infection by MDV occurs, with the release of infectious enveloped virus (Figure 7.3). The FFE infection commences 1–2 weeks after initial infection, and persists for many weeks. Viral antigen and intranuclear inclusion bodies are common in the corneous and transitional layer of the FFE, and affected cells undergo cloudy swelling and hydropic degeneration. The appearance of viral antigen is followed shortly by local perivascular and peri-follicular aggregation and proliferation of lymphoid cells, which in some cases develop into subcutaneous lymphomatous tumours.

Related to infection of FFE are lymphoid lesions that develop in the feather pulp. Moriguchi *et al.* (1982) observed that these pulp lesions resembled those in peripheral nerves, which Fujimoto *et al.* (1971) had earlier classified into

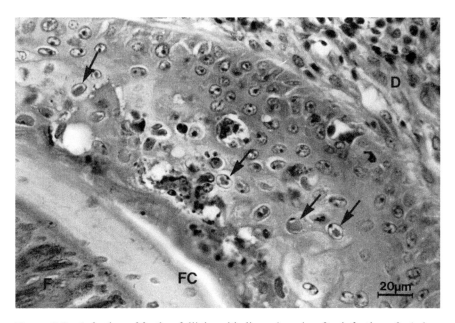

Figure 7.3 Infection of feather follicle epithelium, 4 weeks after infection of a 1-day-old chick with vMDV, showing intranuclear inclusion bodies (arrows) in cells of the transitional layer. D, dermis; F, feather; FC, follicular cavity. (From Payne, 1985, with kind permission.)

two main types: R-type (non-tumourous response) and T-type (tumourous proliferation). Moriguchi *et al.* (1982) classified the pulp lesions into three types: R_1-type – variable infiltration by mainly small lymphocytes; R_2-type – oedema and infiltration by plasma cells, small lymphocytes and heterophils; and T-type – proliferation, usually extensive, by mainly medium lymphocytes or blast cells. Two patterns of pulp lesion development were observed: R_1-type → T-type, occurring in birds that showed persistent inclusion body formation in FFE and lymphoid tumours in viscera, and R_1-type → R_2-type, seen in birds that showed only early transient inclusion body formation in FFE, in general in birds that survived. The use of feather pulp is of particular value in that sequential studies of lymphoid lesions, including tumour progression and regression, may be made without the need to sacrifice the affected bird. More recently, Cho *et al.* (1996) have made sequential studies using skin biopsies of perifollicular cutaneous lesions, which showed two patterns depending on whether or not they were tumour-associated, and related these to pulp lesions.

Infection of other parenchymatous tissues

During the acute infection of lymphoid tissue, and subsequently, other tissues may show evidence of a usually low-grade restrictively productive infection, characterized by cytolysis and often the presence of intranuclear inclusion bodies. This has been observed in adrenal gland, heart, kidney, liver, pancreas and proventriculus.

An immune complex-mediated glomerulopathy has been reported in MD in chickens and quail, with presence of antinuclear antibody (Pradhan *et al.*, 1988; Kaul and Pradhan, 1991).

Peripheral neuropathy

Cellular infiltration of peripheral nerves, leading to their enlargement and paralysis (Figures 7.4, 7.5), is a characteristic feature of classical MD (fowl paralysis). It is often also seen in the more virulent, lymphomatous, form of the disease (acute MD), although when this occurs in adult birds, peripheral nerve lesions may not be readily apparent.

Following infection, the earliest changes in the peripheral nerves are an interneuritic infiltration of mainly macrophages and some lymphocytes, seen under the electron microscope at 5 days (Lawn and Payne, 1979) and under the light microscope at about 10 days (Payne and Rennie, 1973). An accumulation and proliferation of lymphoid cells and macrophages proceeds during the second and third weeks to give rise to the so-called A-type lesion (Payne and Biggs, 1967) (Figure 7.6). This lymphoproliferation has the appearance of a neoplastic lesion, comprised typically of a mixed population of small, medium and large (blastic) lymphocytes, activated reticulum cells and macrophages,

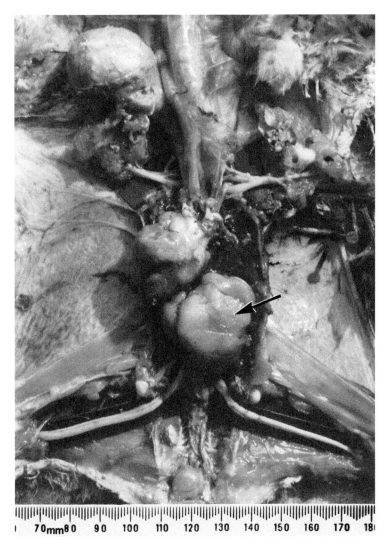

Figure 7.4 Enlarged brachial and sciatic nerves, and an ovarian lymphoma (arrow), in a 5-week-old chicken infected with vMDV. (From Payne, 1985, with kind permission.)

and appears to be identical to the lymphomas that often develop in other tissues (see *Lymphomatosis* below).

The fully developed A-type lesion is accompanied by a primary, segmental, cell-mediated demyelination. In the sequential study of Lawn and Payne (1979), minimal demyelination was observed during the second and third weeks even when severe cellular infiltration was present. Demyelination became marked

Figure 7.5 Paralysis of the legs of a 2-month-old chicken infected with vMDV.

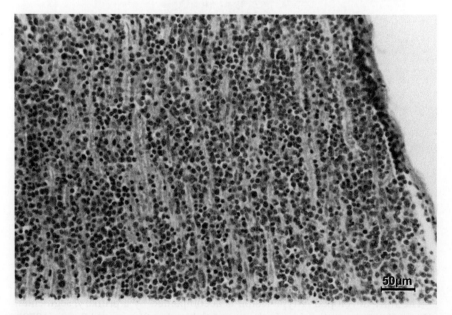

Figure 7.6 A-type nerve lesion, 5 weeks after infection of a 1-day-old chick with vMDV. Note the severe lymphomatous infiltration. (From Payne, 1985, with kind permission.)

during the fourth and fifth weeks, and paralysis became apparent. Ultrastructurally, lymphocytes and macrophages migrated within the basement lamina of the Schwann cells, with destruction of the myelin sheath of nerves. Studies have indicated that the demyelination is mediated immunologically. Skin hypersensitivity reactions to normal myelin, and the presence of tissue-bound and serum antibodies to myelin, have been detected in MD (Schmahl *et al.*, 1975; Pepose *et al.*, 1981). The demyelination was similar to that occurring in experimental allergic neuritis in chickens. The timing of events suggests that the demyelination is secondary to the migration of macrophages and lymphoid cells into peripheral nerve fibres, but the stimulus for this migration has not been elucidated.

Payne and Biggs (1967) described two other types of peripheral nerve lesion. The B-type lesion was characterized by interneuritic oedema and infiltration by small lymphocytes and plasma cells, and demyelination (Figure 7.7). This lesion was considered inflammatory, apparently following an A-type response, and possibly representing a regression of the neoplastic lymphoproliferation. The third, C-type, lesion was characterized by a light infiltration of small lymphocytes and plasma cells, and was considered to be a mild inflammatory lesion. Support for the concept of regression of lymphomatous infiltration in nerves was provided by Burgess *et al.* (2001), who found that B- and C-type neuropathology was observed in a strain of chickens (line 15I) susceptible to MD

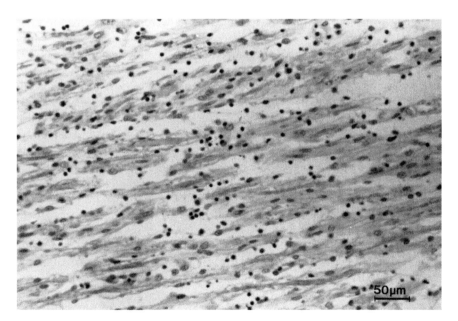

Figure 7.7 B-type nerve lesion, 10 weeks after infection of a 1-day-old chicken with vMDV. Note the interneuritic oedema and sparse cell infiltration. (From Payne, 1985, with kind permission.)

but which showed fluctuating clinical signs of MD or complete recovery. These nerves also contained apoptotic lymphocytes.

In A-type nerves, about 75 per cent of infiltrating lymphocytes are T cells and most of the others are B cells. B and T cells, and plasma cells, are numerous in B-type nerves; in C-type nerves, T cells predominate (Payne and Rennie, 1976b). Burgess *et al.* (2001) found that in the highly susceptible line 7_2 chickens, $CD4^+$ cells predominated over $CD8^+$ cells in A-type lesions, and $CD4^+$ cells also carried the CD30 antigen, considered to be a marker for a transformed cell.

The neuropathology of MD has attracted attention in comparative medicine because of its similarity to the Landry–Guillain-Barre–Strohl syndrome (idiopathic or acute infective polyneuritis) of man (Stevens *et al.*, 1981). Kuroda *et al.* (1989) reported a case of human T-cell leukaemia virus (HTLV)-1 associated adult T-cell leukemia with a demyelinative polyneuropathy in man that was considered to resemble MD, and Vital *et al.* (1990) included an MD-like entity in a new classification of peripheral neuropathy in man.

Central nervous system

Lesions in the central nervous system were long considered unimportant in MD. The most frequent abnormality, seen in the classical and acute form of the disease, was mild perivascular cuffing by mature lymphocytes and macrophages that lie in the perivascular space of Virchow-Robin. Microgliosis and astrocytosis might sometimes be observed. Clinical signs were rarely ascribed to these lesions.

However, in 1959 Zander described a new encephalitic syndrome in adolescent fowl characterized by sudden paralysis, particularly of the neck and legs, usually lasting 24–36 hours, and followed by complete recovery. He termed this syndrome 'transient paralysis' (TP). Kenzy *et al.* (1973) showed that TP could be produced 9–10 days after infection of chickens with an acute MDV, and that it was prevented by vaccination with the herpesvirus of turkeys. The clinical signs were related to a vasculitis and vasogenic oedema, particularly in the cerebellum, with intramyelinic vacuolation and vascular leakage of albumin and immunoglobulin G (Kornegay *et al.*, 1983; Swayne *et al.*, 1989). Perivascular cuffing occurred, but was not considered the cause of the serum protein leakage.

In 1999, Witter and colleagues reported that very virulent plus (vv+) strains of MDV inoculated into 3-week-old chickens caused a TP-like syndrome about 9 days later, but which was rapidly followed by death (Witter *et al.*, 1999). Histologically this new syndrome, which was termed 'acute TP', was similar to the classical TP of Zander, with brain lesions that included acute vasculitis, vasogenic oedema and perivascular cuffing. In studies on central nervous system lesions induced by different pathotypes of MDV, Gimeno *et al.* (1999) identified two further syndromes in addition to classical and acute TP. These were 'persistent neurological disease' (PND) following recovery from TP and associated with lymphoproliferation, and 'late paralysis', the signs of which were similar to TP but which first appeared about 20 days after infection, and were possibly due to infection by contact rather than inoculation.

A detailed analysis of MDV replication, cellular infiltration and MHC antigen expression in the brain following infection with vv+MDV (which induced acute TP) and vMDV (which did not) was made by Gimeno *et al.* (2001a). In MD-susceptible chickens, viral DNA was first detected, by polymerase chain reaction (PCR), at 6 days with vv+MDV and at 8 days with vMDV, and rose to higher levels with the former pathotype. At these time points, endothelial cells appeared to be the first element to be altered. They became hypertrophied, expressed major histocompatibility complex (MHC) class II, and a few apparently expressed the MDV pp38 antigen. However, the question of whether the vasculitis is initiated by infection of endothelial cells is not yet settled. Perivascular infiltration by inflammatory cells (mainly CD8$^+$ T cells and macrophages, with some CD4$^+$ T cells) occurred at the time of appearance of MDV in the brain. Cellularity increased over time, more markedly with vv+MDV, and CD4$^+$ T cells became more frequent. Similar diffuse cellularity occurred in the neuropil. These inflammatory infiltrates were transient, but in vv+MDV-infected brains a proliferation of lymphoblastic CD4$^+$CD8$^-$ T cells appeared from 19 days post-infection. This constituted the later proliferative lesion, which was associated with the PND syndrome. Further findings in this study were that MHC class I expression was down-regulated in many cell types, and MHC class II$^+$ cells increased, during the proliferative phase following vv+MDV infection.

Gimeno *et al.* (2001b) found that there was a differential attenuation of TP and PND induction by MDV, which they presented as a useful tool in the study of pathogenesis.

Lymphomatosis

Multifocal lymphoid proliferation in a variety of tissues begins as early as 1 week after infection, becoming progressively more pronounced and leading to fatal gross lymphomatosis from 3 weeks (Figures 7.4, 7.8). Visceral organs and other sites commonly affected are the gonads, liver, kidneys, spleen, heart, proventriculus, bursa, skeletal muscles and skin. Peripheral nerves are also often affected, as discussed above.

Cytologically the visceral lymphomas are similar in appearance to the infiltration in A-type nerves (Figure 7.9). The lymphoid cells are typically pleomorphic, varying from small lymphocytes to blast cells, although in some cases (particularly in adult birds) blast cells may predominate. About 75 per cent are T cells and most of the others B cells. Although it has been reported that MD lymphomas are monoclonal (Delecluse *et al.*, 1993), recent work refutes this (Burgess and Davison, 2002).

Lymphoblastoid cells lines established from lymphomas almost always have the CD4$^+$CD8$^-$ (helper T cell) phenotype, and this has been considered the usual neoplastic component of the lymphoma. However, lymphoblastoid cell

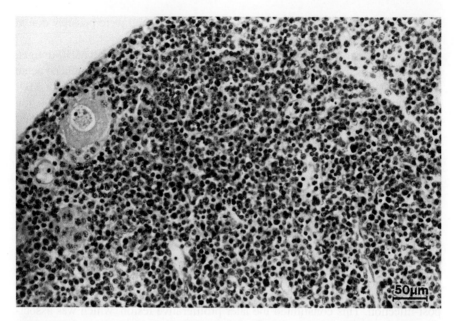

Figure 7.8 Ovarian lymphoma, same case as Figure 7.5. (From Payne, 1985, with kind permission.)

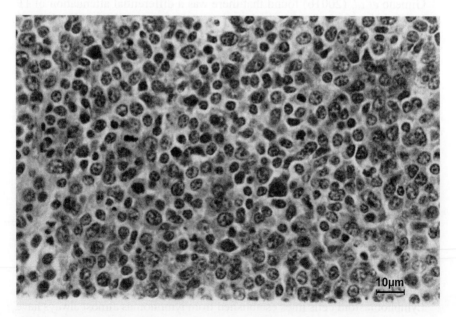

Figure 7.9 Ovarian lymphoma, showing varying morphology of lymphoid cells; same case as Figure 7.8. (From Payne, 1985, with kind permission.)

lines derived from the 'local lymphoid lesion' model (resulting from the inoculation of MDV-infected allogeneic cells into the wingweb or breast muscle of young chicks) had a multiplicity of phenotypes, including $CD8^+$ cells (Calnek et al., 1989; Schat et al., 1991), indicating that susceptibility to neoplastic transformation is not restricted to a specific cell type.

MD lymphoblastoid cell lines are non-productively infected, with multiple copies of MDV genome, are essentially diploid, and carry a variety of antigenic markers – including chicken foetal antigen, Forssman antigen, *Ia*-like antigen and MHC antigens. Another antigen, termed MATSA, was originally considered to be an MD tumour-specific antigen (Powell et al., 1974) or a MD tumour-associated surface antigen (Witter et al., 1975), but other studies (see Burgess and Nair, 2002) suggest that a variety of antigens have subsequently been covered by the term 'MATSA', including an antigen on non-transformed, activated T cells (McColl et al., 1987). Whether MD tumour-specific antigens exist bears on the question of whether anti-tumour immunity as well as anti-viral immunity is involved in resistance to MD. Moore et al. (1993) found a specific chromosomal aberration affecting the short arm of a chromosome 1 homologue in 14/15 MD lymphoblastoid cell lines, but its significance in MD transformation is still uncertain.

Recently, another antigen recognized by the AV37 monoclonal antibody has been associated with MDV transformed cells and appears to be the avian homologue of CD30 that is greatly over-expressed in MD tumour cells (Burgess and Davison, 2002). These authors have questioned whether MD lymphoblastoid cell lines are representative of neoplastically transformed cells found in MD lymphomas *in vivo*, and they provide evidence that a lymphoid cell highly expressing the CD30 antigen ($CD30^{hi}$) is the transformed cell in lymphomas.

Leaving aside these debates, the MD lymphoma is clearly cytologically complex, consisting of neoplastically transformed T cells, and a variety of other cells (some $CD4^+$ and $CD8^+$ T cells, B cells, macrophages, and natural killer cells) that are involved in immunological reactions against transformed cells (Sharma, 1983; Burgess and Davison, 2002). Recent work suggests also that there is a structure to the development of lymphomas in terms of transformed and reactive cells (Burgess et al., 2001). For further information and discussion on the nature of MD tumours, see Chapter 8.

Atherosclerosis

An association between coronary atherosclerosis and neurolymphomatosis (Marek's disease) was first observed by Patterson and Cottrall in 1950 (Patterson and Cottrall, 1950). The association was investigated by Fabricant and colleagues (see Fabricant, 1985; Fabricant and Fabricant, 1988), who found that mild atherosclerotic arterial lesions found in chickens without MDV infection were greatly increased in severity in the presence of infection by a mildly pathogenic MDV and supplemental cholesterol in the diet. The lesions were classified as fatty, proliferative, and fatty-proliferative. MDV antigens were detected by immunofluorescence in the medial layer of affected arteries, apparently in smooth muscle cells, suggesting that virus-induced medial necrosis is an initial change in the development of atherosclerosis. MDV infection affected enzymatic activities in arterial smooth muscle cells *in vitro* and *in vivo*, resulting in increased concentrations of cellular cholesterol and cholesteryl ester (Hajjar *et al.*, 1985, 1986). Endothelial hyperplasia was observed in arteries and veins of MDV-infected birds. More recently, Kariuki Njenga and Dangler (1995) observed expression of MHC class II antigen on vascular endothelium 2 weeks after MDV infection, and suggested that an immunopathogenic mechanism might be involved in the early pathogenesis of atherosclerosis.

Atherosclerosis in chickens shows similarities to the disease in man, and the involvement of a herpesvirus in the avian disease has attracted interest because another herpesvirus, cytomegalovirus, has been associated with the human disease.

Ocular changes

Early workers described so-called ocular lymphomatosis, considered by many to be a form of MD. Common findings are lymphocytic, plasma cell and sometimes heterophil infiltration of the iris, ciliary body and conjunctiva, and less commonly of the choroid membrane, pecten and retina. Many reports were based on field cases, but similar lesions have been produced experimentally (Smith *et al.*, 1974).

More recently, MD outbreaks have been reported in which blindness was a common feature, with extensive infiltration by inflammatory cells in the cornea, uveal tract, retina and pecten, intranuclear inclusion bodies in mononuclear cells and retinal cells, and degeneration of photoreceptors and other retinal cells (Ficken *et al.*, 1991; Spencer *et al.*, 1992). Mild paralysis and depression were observed, but the incidence of visceral tumours was low. The isolated viruses had a tropism for ocular tissues, suggesting that they represent a new pathotype.

Factors affecting lesion type

Virus strain

From the time that different strains of MDV were isolated in the 1960s, it has been recognized that they vary in their oncogenicity and virulence. The current and commonly used classification of MDV pathotypes derives particularly from the work of Witter (1997), who has recognized the continuing evolution and selection of strains of increasing virulence apparently as a consequence of the development and use of new MD vaccines. All virulent or oncogenic strains of MDV belong to serotype 1, and currently four pathotypes are recognized, as follows:

(a) mMDV (mild MDV) – in genetically susceptible chickens these strains cause mainly neural MD and sometimes a low incidence of mainly ovarian lymphomas; their pathogenic effects are preventable with HVT vaccines.

(b) vMDV (virulent MDV) – these cause a high incidence of visceral and neural lymphomas; their effects are prevented by HVT vaccines.

(c) vvMDV (very virulent MDV) – these cause a high incidence of visceral and neural lymphomas. They are oncogenic in HVT vaccinated birds and in birds genetically resistant to less virulent viruses; their effects are preventable with bivalent vaccines.

(d) vv+MDV (very virulent plus MDV) – these cause a high incidence of lymphomas and are oncogenic in birds vaccinated with bivalent vaccines.

This classification is based on the ability of emerging strains of MDV to overcome different vaccination strategies using a standard susceptible strain of chickens. Increasing virulence in this sense appears to be related to increasing oncogenicity, but the molecular basis and the mechanisms of viral virulence and oncogenicity are not well understood. The more virulent strains of MDV also appear to be those that cause more severe acute cytolytic disease and acute TP, and MDV-induced immunosuppression may also be a factor. Lymphoma formation in adult birds also appears to be associated with increased virulence.

A new system for classifying MDV based on neuropathotyping (of brain lesions) was developed by Gimeno *et al.* (2002), but was considered to complement rather than replace conventional pathotyping. (The disadvantage of the conventional pathotyping method is that it is dependent on the use of chicken strains not readily available to all laboratories, so that it is not possible to confidently categorize every new MDV isolate into one of the four pathotypes.)

Host

The genetic constitution of the host plays an important part in determining the outcome of infection by MDV (see Chapter 10). The best understood genetic locus influencing resistance to MD is the B-F region of the MHC B-locus in chickens, closely linked to B-G genes encoding B blood group antigens. The B^{21} blood group allele is especially notable for its association with MD resistance that is absent at hatching but develops in early life, as exemplified by the Cornell N line. It is expressed as reduced numbers of infected T cells compared with susceptible birds, and may be due to a superior immunological ability to reject infected or transformed T cells. The early cytolytic lymphoid infection occurs, but lymphomas do not develop. This type of genetic resistance is probably synonymous with so-called age resistance.

A second type of genetic resistance is associated with the non-MHC MDV-1 locus. It is present at hatching, and is associated with reduced early cytolytic lymphoid infection and reduced numbers or susceptibility of target T cells for MDV infection or transformation, as exemplified by RPL line 6_1. Baigent and Davison (1999) studied the splenic lymphoid cell response of this line to vMDV infection. In comparison with the MHC-matched susceptible line 7_2, they observed fewer infected B cells and an early but limited increased number of CD4 and CD8 TCRαβ$^+$ cells thought to limit cells entering latent infection.

Studies on CD30^{hi} leukocytosis (see under *Lymphomatosis*) have revealed the presence of these putatively neoplastic transformed cells both early and late after MDV infection in genetically resistant as well as genetically susceptible birds (Burgess and Davison, 2002). In this scenario, it is suggested that lymphoma development in susceptible birds is a consequence of evasion, or absence, of cell-mediated immunity to transformed cells.

Acquired immunity also influences susceptibility to MD. This includes passive immunity due to maternal antibody, which lasts for about 3 weeks, and actively acquired immunity because of either natural infection by MDV of low virulence or MD vaccination, which is of long or permanent duration. These forms of immunity suppress the early cytolytic infection and lymphoproliferation caused by virulent virus.

Marek's disease in other species

MD is primarily a disease of chickens, but turkeys, quail and pheasants can also be affected. Other species of birds are not recognized as being susceptible.

Turkeys

Outbreaks of MD in commercial turkeys have occurred in France, Germany, Israel and the UK over the past 10 years or so (Pennycott and Venugopal, 2002), and the disease can be induced experimentally in turkeys (Witter *et al.*, 1974; Paul *et al.*, 1977). Of three strains of vMDV, MD mortality in turkeys was induced only with the GA strain (Powell *et al.*, 1984). Visceral lymphomas were the main lesions observed in experimentally infected turkeys, and they appeared to be less susceptible than chickens to early cytolytic infection, early virus replication and peripheral nerve lesions. Lymphoblastoid cell lines established from turkey MD were reported to be B cells by Nazerian *et al.* (1982) but T cells by Powell *et al.* (1984) (see also Chapter 10). Remarkably, vaccination with HVT did not prevent MD in turkeys (Elmubarak *et al.*, 1982), but CVI 988 (Rispens) vaccine has been found to be effective in the field.

Japanese quail

Outbreaks of MD occur in Japanese quail in countries such as India and Japan, where this species is raised commercially (Pradhan *et al.*, 1985; Koboyashi *et al.*, 1986), and MD has been induced experimentally in quail with isolates of MDV from chickens (Dutton *et al.*, 1973; Mikami *et al.*, 1975). Lymphomas occur in various organs, but peripheral nerves are rarely involved. HVT is used for commercial vaccination (Koboyashi *et al.*, 1986). Recently, MD has been reported in commercially raised Japanese quail in Scotland (Pennycott *et al.*, 2003).

Pheasants

MD-like lymphoma and nerve lesions have been reported in pheasants, and can be reproduced experimentally with MDV. However, MD does not appear to be common in commercially reared pheasants, and vaccination is not used.

Conclusions

Pathology has long been recognized as the core of medicine (human and veterinary), providing a framework for the understanding, treatment and prevention of diseases. It has continually evolved, from morbid anatomy, histopathology and ultrastructural pathology to the so-called new pathology based on cellular phenotypes and genotypes and on molecular (nucleic acid) pathology. In Marek's disease research, deeper understanding of the disease processes that these modern approaches provide has been slow to arrive, due mainly to a research emphasis on the virus and vaccines and the limited resources for the development and application of the newer techniques.

However, the past 10 or so years have seen the identification of new chicken cell markers and reagents for their detection, and the development of molecular techniques for detection and localization of MD virus and antigens. These new tools have been successfully used to further understanding of several pathological processes in MD, such as early inflammatory changes in lymphoid tissues, lymphomagenesis, and brain lesions, as described in this chapter. Other lesions and processes await their application, such as the development of early peripheral nerve lesions and early visceral lymphoid lesions. Work needs to continue on the nature of the neoplastically transformed cell, the cytology and nature of lymphomas, the possible 'progression' of lymphoid tumour cells, peripheral nerve demyelination, and vascular changes. Moreover, the continuing appearance of new strains of MDV with changed virulence and tropisms, and the increase of MD in species other than chickens, suggest no shortage of pathological topics needing investigation.

References

Baigent, S. J. and Davison, T. F. (1999). *Avian Pathol.*, **28**, 287–300.

Barrow, A. and Venugopal, K. (1999). *Acta Virol.*, **43**, 90–93.

Barrow, A. D., Burgess, S. C., Howes, K. and Nair, V. K. (2003). *Avian Pathol.*, **32**, 183–191.

Burgess, S. C. and Davison, T. F. (2002). *J. Virol.*, **76**, 7276–7292.

Burgess, S. C. and Nair, V. K. (2002). In *Modern Concepts of Immunology in Veterinary Medicine – Poultry Immunology* (ed. T. Mathew), pp. 243–300. Thajema Publishing, West Orange.

Burgess, S. C., Basaran, B. H. and Davison, T. F. (2001). *Vet. Pathol.*, **38**, 129–142.

Calnek, B. W. (2000). In *Marek's Disease* (ed. K. Hirai), pp. 25–55. Springer-Verlag, Berlin.

Calnek, B. W., Lucio, B., Schat, K. A. and Lillehoj, H. S. (1989). *Avian Dis.*, **33**, 291–302.

Cho, K. O., Mubarak, M., Kimura, T. *et al.* (1996). *Avian Pathol.*, **25**, 325–343.

Delecluse, H. J., Schüller, S. and Hammerschmidt, W. (1993). *EMBO J.*, **12**, 3277–3286.

Dutton, R. L., Kenzy, S. G. and Becker, W. A. (1973). *Poultry Sci.*, **52**, 139–143.

Elmubarak, A. K., Sharma, B. D., Witter, R. L. and Sanger, V. L. (1982). *Am. J. Vet. Res.*, **43**, 740–742.

Fabricant, C. G. (1985). *Adv. Vet. Sci. Comp. Med.*, **30**, 39–66.

Fabricant, C. G. and Fabricant, J. (1988). *Proceedings of the 3rd International Symposium on Marek's Disease, Osaka*, pp. 317–323, Japanese Association on Marek's Disease, Japanese Association on Marek's Disease, Osaka.

Ficken, M. D., Nasisse, M. P., Boggan, G. D. *et al.* (1991). *Avian Pathol.*, **20**, 461–474.

Fujimoto, Y., Nakagawa, M., Okada, K. *et al.* (1971). *Jpn J. Vet. Res.*, **19**, 7–26.

Gilka, F. and Spencer, J. L. (1995). *Avian Pathol.*, **24**, 393–410.

Gimeno, I. M., Witter, R. L. and Reed, W. M. (1999). *Avian Dis.*, **43**, 721–737.

Gimeno, I. M., Witter, R. L., Hunt, H. D. *et al.* (2001a). *Vet. Pathol.*, **38**, 491–503.

Gimeno, I. M., Witter, R. L., Hunt, H. D. *et al.* (2001b). *Avian Pathol.*, **30**, 397–409.

Gimeno, I. M., Witter, R. L. and Neumann, U. (2002). *Avian Dis.*, **46**, 909–918.

Hajjar, D. P., Falcone, D. J., Fabricant, C. and Fabricant, J. (1985). *J. Biol. Chem.*, **260**, 6124–6128.

Hajjar, D. P., Fabricant, C. G., Minick, C. R. and Fabricant, J. (1986). *Am. J. Pathol.*, **122**, 62–70.

Kariuki Njenga, M. and Dangler, C. A. (1995). *Vet. Pathol.*, **32**, 403–411.

Kaul, L. and Pradhan, H. K. (1991). *Vet. Immunol. Immunopathol.*, **28**, 89 96.

Kenzy, S. G., Cho, B. R. and Kim, Y. (1973). *J. Natl Cancer Inst.*, **51**, 977–982.

Koboyashi, S., Koboyashi, K. and Mikami, T. (1986). *Avian Dis.*, **30**, 816–819.

Kornegay, J. N., Gorgacz, E. J., Parker, M. A. *et al.* (1983). *Am. J. Vet. Res.*, **44**, 1541–1544.

Kuroda, Y., Nakata, H., Kakigi, R. *et al.* (1989). *Neurology*, **39**, 144–146.

Lawn, A. M. and Payne, L. N. (1979). *Neuropathol. Appl. Neurobiol.*, **5**, 485–497.

McColl, K. A., Calnek, B. W., Harris, W. V. *et al.* (1987). *J. Natl Cancer Inst.*, **79**, 991–1000.

Mikami, T., Onuma, M., Hayashi, T. *et al.* (1975). *J. Natl Cancer Inst.*, **54**, 607–614.

Moore, F. R., Schat, K. A., Hutchison, N. *et al.* (1993). *Intl J. Cancer*, **54**, 685–692.

Moriguchi, R., Fujimoto, Y. and Izawa, H. (1982). *Avian Dis.*, **26**, 375–388.

Morimura, T., Hattori, M., Ohashi, K. *et al.* (1995*). J. Gen. Virol.*, **76**, 2979–2985.

Morimura, T., Ohashi, K., Kon, Y. *et al.* (1996). *Arch. Virol.*, **141**, 2243–2249.

Nazerian, K., Elmubarak, A. K. and Sharma, J. M. (1982). *Intl J. Cancer*, **29**, 63–68.

Patterson, J. C. and Cottrall, G. E. (1950). *Arch. Pathol.*, **49**, 699–709.

Paul, P. S., Sautter, J. H. and Pomeroy, B. S. (1977). *Am. J. Vet. Res.*, **38**, 1653–1656.

Payne, L. N. (1985). In *Marek's Disease* (ed. L. N. Payne), pp. 43–75. Martinus Nijhoff Publishing, Boston.

Payne, L. N. and Biggs, P. M. (1967). *J. Natl Cancer Inst.*, **39**, 281–302.

Payne, L. N. and Rennie, M. (1973). *J. Natl Cancer Inst.*, **51**, 1559–1573.

Payne, L. N. and Rennie, M. (1976a). *Intl J. Cancer*, **18**, 510–520.

Payne, L. N. and Rennie, M. (1976b). *Avian Pathol.*, **5**, 147–154.

Payne, L. N., Frazier, J. A. and Powell, P. C. (1976). *Intl Rev. Exp. Pathol.*, **16**, 59–153.

Pennycott, T. W. and Venugopal, K. (2002). *Vet. Rec.*, **150**, 277–279.

Pennycott, T. W., Duncan, G. and Venugopal, K. (2003). *Vet. Rec.*, **153**, 293–297.

Pepose, J. S., Stevens, J. G., Cook, M. L. and Lampert, P. W. (1981). *Am. J. Pathol.*, **103**, 309–320.

Powell, P. C., Payne, L. N., Frazier, J. A. and Rennie, M. (1974). *Nature*, **251**, 79–80.

Powell, P. C., Howes, K., Lawn, A. M. *et al.* (1984). *Avian Pathol.*, **13**, 201–214.

Pradhan, H. K., Mohanty, G. C. and Mukit, A. (1985). *Avian Dis.*, **29**, 575–582.

Pradhan, H. K., Mohanty, G. C., Lee, W. Y. *et al.* (1988). *Vet. Immunol. Immunopathol.*, **19**, 165–171.

Schat, K. A. (1987). *Cancer Surveys*, **6**, 1–37.

Schat, K. A., Chen, C. L. H., Calnek, B. W. and Char, D. (1991). *J. Virol.*, **65**, 1408–1413.

Schmal, W., Hoffmann-Fezer, G. and Hoffmann, R. (1975). *Z. Immunitaetsforsch. Exp. Klin. Immunol.*, **150**, 175–183.

Sharma, J. M. (1983). *Vet. Immunol. Immunopathol.*, **5**, 125–140.

Shek, W. R., Calnek, B. R., Schat, K. A. and Chen, C. L. H. (1983). *J. Natl Cancer Inst.*, **70**, 485–491.

Smith, T. W., Albert, D. M., Robinson, N. *et al.* (1974). *Invest. Ophthalmol.*, **13**, 586–592.

Spencer, J. L., Gilka, F., Gavora, J. S. *et al.* (1992). *Proceedings of the 4th International Symposium on Marek's Disease, Amsterdam*, pp. 199–201, WPSA, Amsterdam.

Stevens, J. G., Pepose, J. S. and Cook, M. L. (1981). *Ann. Neurol.*, Suppl. **9**, 102–106.

Swayne, D. E., Fletcher, O. J. and Schierman, L. W. (1989). *Avian Pathol.*, **18**, 385–396.

Vital, C., Vital, A., Julien, J. *et al.* (1990). *J. Neurol.*, **237**, 177–185.

Witter, R. L. (1997). *Avian Dis.*, **41**, 149–163.

Witter, R. L. and Schat, K. A. (2003). In *Diseases of Poultry* (ed. Y. M. Saif), 11th edn, pp. 407–465. Iowa State Press, Ames.

Witter, R. L., Solomon, J. J. and Sharma, J. M. (1974). *Am. J. Vet. Res.*, **35**, 1325–1332.

Witter, R. L., Stephens, E. A., Sharma, J. M. and Nazerian, K. (1975). *J. Immunol.*, **115**, 177–183.

Witter, R. L., Sharma, J. M. and Fadly, A. M. (1980). *Avian Dis.*, **24**, 210–232

Witter, R. L., Gimeno, I. M., Reed, W. M. and Bacon, L. D. (1999). *Avian Dis.*, **43**, 704–720.

Marek's disease lymphomas

8

SHANE C. BURGESS

Department of Basic Sciences, College of Veterinary Medicine,
Mississippi State University, Mississippi, USA

Introduction

MD was initially described by József Marek (Marek, 1907) as a 'polyneuritis', i.e. a cellular inflammation of peripheral nerves. Later, the neoplastic nature of MD was recognized (Pappenheimer *et al.*, 1926, 1929). Two forms of lymphomatous MD were then defined: 'classical' and 'acute' (see Chapter 2). In classical MD, gross lymphomas predominate in the peripheral nerves. 'Acute' MD describes the lymphoid necrosis and lymphomas of many visceral organ tissues (Benton and Cover, 1957; Biggs *et al.*, 1965) that began to occur after more virulent MD virus (MDV) emerged. The clinical signs of MD can be attributed to the nature of MD lymphoma, which causes organ dysfunction. In classical MD, lymphomas of the brachial and lumbo-sacral neural plexuses cause paralysis. Nerves increase to up to three times their normal diameter and appear grey/yellow, rubbery and oedematous. The visceral lymphomas typical of acute MD commonly occur in the adrenal gland, heart, intestine, kidney, liver, lung, ovary, proventriculus, serosa, skeletal muscle, spleen and thymus. MD lymphomas in these organs are clearly visible to the naked eye. The skin, specifically the feather-follicle epithelium (FFE), is associated with MDV shedding and is also a site for lymphomagenesis. In other words, the two components of so-called 'skin leukosis' are FFE cytolysis and dermal lymphocytic infiltration. Occasionally there is iridocyclitis, with or without involvement of the periorbital tissues, including the Harderian gland. The frequency and severity to which each organ is affected varies, depending on the chicken genotype and MDV strain (reviewed by Payne, 1985).

Histopathology

Regardless of their anatomical location, MD lesions have common fundamental histopathological features. Early and/or mild pathology is distinguished by perivascular mononuclear cell 'cuffing' from 5–7 days post-infection (dpi) – during the cytolytic phase of infection – suggesting both increased vascular permeability and leukocyte adhesion with extravasation (reviewed by Payne, 1985). Most MD lymphoma cells contain the MDV genome (Ross *et al.*, 1981), but very few produce MDV virions – i.e. most lymphoma cells are latently infected. MD lymphomas are cytologically complex (reviewed Payne, 1985). Most MD lymphoma cells are T lymphocytes (Payne and Rennie, 1976; Hoffmann-Fezer and Hoffmann, 1980) that are not neoplastically transformed (Sharma, 1981). A minority (1 per cent to ~35 per cent) of T lymphocytes express molecules known as 'MD-associated tumour surface antigen' (MATSA), thought to be markers of neoplastic-transformation (reviewed Burgess and Venugopal, 2002). B lymphocytes, natural killer cells, macrophages and other non-lymphoid cells are a minority in MD lymphomas, and these are not neoplastically transformed (reviewed Payne, 1985). Ultrastructurally, MD lymphoma cells cannot be distinguished from normal lymphocytes or lymphoblasts (Doak *et al.*, 1973; Payne *et al.*, 1976).

Peripheral nerves

Two main peripheral nerve pathologies are described; neoplastic proliferation (with or without secondary demyelination) and primary cell-mediated demyelination (Wight, 1969; Prineas and Wright, 1972; Lampert *et al.*, 1977). The pathology of peripheral nerves and the central nervous system is fully described in Chapter 7.

Skin

The FFE is the only location from which fully infectious MDV is shed into the environment. Given the lymphotropic nature of MDV, it is logical that lymphoid infiltration is necessary to transport MDV from its site of infection to its site of egress. Skin pathology has two components; lytic infection of the FFE and perivascular accumulation of lymphocytes in the dermis. Perivascular lymphocyte accumulation occurs from 5–7 dpi. If these lymphocyte accumulations increase in size and become visible to the naked eye, they are termed 'skin leukosis' (reviewed by Payne, 1985).

Visceral and skeletal muscle lymphomas

Lymphomas consist predominantly of heterogeneous accumulations of T lymphocytes and lymphoblasts; B lymphocytes, macrophages (Plate 3A) and reticulum cells are present as minorities. Microscopic multifocal lymphomas are present ~5–7 dpi and, depending on the MDV pathotype and chicken geno-type, these can become grossly visible from ~14 dpi (Payne and Rennie, 1976; Burgess *et al.*, 2001).

The general paradigm for MD lymphoma structure described in 1985 has since been validated with many chicken genotypes and MDV pathotypes, including 'hypervirulent' MDV pathotypes (Burgess *et al.*, 2001, 2002; Gimeno *et al.*, 2001). Hypervirulent MDV pathotypes may be so oncogenic that the lymphomas they produce appear to outgrow their blood supply (Parcells and Burgess, 2004). The implication is that MDV evolution is not just towards increased cytolysis, but also towards more rapid lymphomagenesis. If faster lymphomagenesis is occurring, it reinforces the point that understanding MD lymphomagenesis is critical to maintaining control over MDV. Furthermore, faster lymphomagenesis suggests that selection imposed by current vaccination strategies occurs at both the cytolytic and latent/transformation phases.

Lymphoma cell phenotypes

Classical histopathology and the earliest immunohistochemical studies using polyclonal antisera described the general structure of MD lymphomas but only gave hints to the possible structure–function relationships within lymphomas. The general approach taken has been to describe MD lymphoma cells in terms of the MDV genome and the molecules expressed by the cells. This section will summarize what is known about MDV infection, MDV gene and host gene expression in MD lymphoma cells. It will discuss the role of Marek's-associated tumour surface antigens (MATSAs), MDV-transformed chicken cell (MDCC) cultures and phenotyping of lymphoma cells directly *ex-vivo* and *in-situ*.

MATSAs

No discussion of MD lymphoma cell phenotyping can exclude MATSAs. Although Powell *et al.* (1974) suggested that the recently developed MDCC expressed a 'membrane tumour-specific antigen', the term MATSA originates from Witter *et al.* (1975). MATSAs are a diverse group of surface antigens expressed by MD lymphoma cells. MATSAs have been detected using poly-clonal sera and monoclonal antibodies (mAbs); their physico-chemical proper-ties are variably defined. The MATSAs are almost certainly only related in as much as they are expressed on the surface of MD lymphoma cells (reviewed by Burgess and Venugopal, 2002). Some could even be MDV antigens. To date, only one molecule that can be classified as a MATSA has been definitively identified. This molecule is discussed in detail below.

Contemporary oncology defines tumours by quantitative gene expression at both the messenger (m)RNA and protein levels (Liotta and Petricoin, 2000).

Surface-expressed tumour antigens are particularly useful for identifying neoplastically transformed cells. The antibodies used to recognize the MATSAs have been used to identify neoplastically transformed cells in MD-lymphomas (Payne, 1985), and also as diagnostic tools (Sharma, 1985). Interestingly, MATSA-expressing cells occur in both MD-resistant and MD-susceptible chickens (Murthy and Calnek, 1979), and one MATSA has been detected on lymphoblastoid cells in the nerves of chickens and turkeys vaccinated with HVT (Calnek *et al.*, 1979; Fabricant *et al.*, 1982).

The neoplastically transformed state of cells expressing MATSAs has never been proven. The association of some of the MATSAs with neoplastic transformation has been questioned because some of the antibodies identified: (1) foetal cells (Murthy *et al.*, 1979); (2) lymphocytes infected with HVT and serotype 2 (Schat and Calnek, 1978); and (3) activated lymphocytes (McColl *et al.*, 1987). In the light of recent knowledge concerning tumour antigens, none of these reasons should exclude any of the MATSAs as tumour-antigens (Burgess and Venugopal, 2002).

MDV-transformed cells

A logical approach to identifying the phenotype of neoplastically transformed MD cells was to culture them *in vitro*, directly from lymphomas. This approach was used to identify the first oncogenes of acutely transforming retroviruses. The theory was that the neoplastically transformed MD lymphoma cells should be 'immortalized' and therefore should replicate in a mixed MD lymphoma cell culture, while non-transformed MD lymphoma cells should die. We now know that MDV has no classical acutely transforming oncogene *per se* (see Chapter 4), and that the lymphoma environment is critical for neoplastically transformed cells (Burgess and Davison, 2002). This perhaps explains the difficulty in producing the first MDCC lines (Akiyama *et al.*, 1973; Powell *et al.*, 1974).

The surface antigens known to be expressed by MDCC lines have recently been reviewed (Parcells and Burgess, 2004). Most MDCC examined have phenotypes associated with activated (tissue-invasive) CD4$^+$ T-helper lymphoblasts. Although most MDCC have been produced from natural lymphomas, a model MDV-transformation system using MDV-infected allogeneic chicken kidney cells injected into the wingweb or muscle has been used to produce MDCC lines (Calnek *et al.*, 1989). After injection, sections of the lesions were excised and the infiltrating lymphocytes cultured to generate cell lines that were either CD4$^+$, CD8$^+$, CD4$^+$CD8$^+$ or CD4$^-$CD8$^-$ (Schat *et al.*, 1991).

Phenotypes of MD lymphoma cells

Four major techniques have been used to identify MDV infection as well as MDV gene and host-gene expression in MD lymphoma cells. These techniques are *in situ* hybridization (ISH) and the polymerase chain reaction (PCR), both

of which recognize the MDV DNA genome and mRNAs, while immunohisto-
chemistry and flow cytometry have been used to identify antigens.

MDV genome

Ross *et al.* (1981) reported that almost all of the T lymphocytes in MD lymphomas contain the MDV genome, a small amount of which is transcribed. This
suggests first that MDV must be present in a latent state (because MDV productive infection is cytolytic), and secondly that latency involves mRNA transcription. Thirdly, because (as stated above) only a small minority of cells in
MD lymphomas are neoplastically transformed and the rest are 'reactive' T
lymphocytes, the presence of latent MDV in T lymphocytes is not in itself
indicative of neoplastic transformation.

To examine the structure of the MDV genome in MD lymphoma cells and its
relationship to the host DNA, ISH was performed using metaphase chromosomes isolated from MD lymphoma cells (Delecluse and Hammerschmidt,
1993). This work describes that MDV integration into multiple chromosomal
sites is a random event; that circular MDV genomes (episomes) are absent; and
that linear virus DNA is rare. The integration pattern led the authors to propose
that MD lymphomas are clonal.

Moore *et al.* (1993) examined MDCC chromosome structure directly. An
extra G-positive band and interband on the short arm of one chromosome I
homologue was observed in 14 out of 15 MDCC cultures examined. Amplified
genomic DNA sequences were identified by ISH that linked to an endogenous
retrovirus locus and genes in the histone multi-gene family. Bloom (1970) suggested the chromosomal aberration was an essential element of MD neoplastic
transformation and part of the multi-step oncogenic process. The work was
done using MDCC cultures; unfortunately the hypothesis presented has never
been tested directly in MD lymphoma cells.

MDV genome expression is still being investigated extensively (reviewed by
Morgan *et al.*, 2001; Parcells and Burgess, 2004). MDV genome expression
work mainly involves defining the transcriptional and translational activity of
the MDV genome in latency and when MDV is reactivating from latency. It
is performed mainly using MDCC cultures. A review of this work is given in
Chapter 4. Little work has specifically addressed the issue of MDV gene transcription in lymphomas.

The cellular environment

Direct analysis of MD lymphoma antigens in tissue sections is laborious, expensive and time-consuming, mainly because it requires live chickens. Lesions after
MDV infection have been described as either 'inflammatory' or 'proliferative'
(Payne and Biggs, 1967). Until the mid-1980s, analysis was based on the histopathological appearance of cells in lesions/lymphomas. Since then, immunostaining methods have been used to define populations of cells in MD lymphoma.

The general kinetics of MD lesion/lymphoma formation, regardless of tissue type or infection route, are that from ~3–5 dpi monocytes are the first cells to invade the tissues (Payne and Biggs, 1967; Burgess *et al.*, 2001; Gimeno *et al.*, 2001; Barrow *et al.*, 2003). From ~5–7 dpi a lymphoid infiltration is present with approximately equal proportions of CD4$^+$ T-helper lymphocytes and CD8$^+$ T-killer lymphocytes. What happens next critically depends on two inter-acting variables: the chicken's genotype and the MDV pathotype. If chickens are MD-susceptible relative to the virulence of the particular MDV, the CD4$^+$ lymphocyte numbers increase and the CD8$^+$ lymphocyte numbers decrease, and gross CD4$^+$ lymphomas form (Plate 3A; Burgess *et al.*, 2001). The oppos-ite occurs if chickens are resistant relative to the virulence of the particular MDV; CD8$^+$ lymphocytes predominate over CD4$^+$ lymphocytes, and lesions regress (Burgess *et al.*, 2001; Gimeno *et al.*, 2001).

Clinical syndromes peculiar to the central nervous system (CNS) can be divided into two categories (see Chapter 7): those that are transient, mild and non-lethal; and those that are acutely lethal. In the transient disease, monocytes first invade the CNS followed by CD4$^+$ and CD8$^+$ lymphocyte infiltrations. CD8$^+$ lymphocytes then predominate, and the lesions resolve. Such CNS lesions have been described as 'inflammatory' (Gimeno *et al.*, 2001). This pathology resembles non-CNS regressing lesions (Burgess *et al.*, 2001).

In the case of extremely virulent MDV (Gimeno *et al.*, 1999, 2001), monocytes first invade the CNS, followed by CD4$^+$ and CD8$^+$ lymphocytes. If CD4$^+$ lympho-cytes predominate, then 'proliferative' lesions form. This type of CNS pathology resembles non-CNS lesions that progress to frank lymphomas (Burgess *et al.*, 2001). Unlike the equivalent non-CNS lesions, death occurs before gross lymph-omas form (Gimeno *et al.*, 1999, 2001). The difference is possibly a function of lower functional buffering capacity of the CNS to sustain proliferative lesions.

The similarities in cellular profiles in both 'inflammatory' and 'proliferative' (or 'progressive' and 'regressive') lesions between CNS and non-CNS disease suggest they are different clinical manifestations of the same mechanisms of molecular pathogenesis. This is reinforced because both non-CNS and CNS lesions have clusters of CD4$^+$ cells that express the MDV oncoprotein Meq within the larger CD4$^+$-predominant proliferating lesions (Burgess *et al.*, 2001; Gimeno *et al.*, 2001). Based on the expression of Meq, Gimeno *et al.* (2001) suggested that neoplastically-transformed cells are present in those brain lesions caused by the hypervirulent MDV. This suggestion is supported by the work using the AV37 mAb described in the next section.

The identity of the neoplastically transformed cells in MD lymphomas

During work at the IAH in the early 1990s, a panel of mAbs was raised against MDCC-HP9 (T. F. Davison, personal communication). One of these mAbs,

AV37, has been of critical importance in investigations on MD lymphoma structure.

AV37 has been used to analyse non-CNS MD lymphomas by immunohisto-chemistry, flow cytometry, and in magnetically-activated cell sorting to isolate cells for further analysis (Burgess and Davison, 1999, 2002; Burgess et al., 2001). AV37 identifies clusters of cells in MD lesions/lymphomas of both MD-resistant and MD-susceptible inbred and outbred chickens, as well as turkeys (Burgess and Davison, 1999, 2002; Burgess et al., 2001; I. M. Gimeno, personal communication).

AV37 recognizes the CD30 antigen, a member of the tumour necrosis factor receptor II family (Burgess et al., 2004). The neoplastically transformed cells in MD lymphomas, like their human cousins, over-express CD30 (CD30^{hi}: Burgess and Davison, 2002). It is critical to emphasize that it is gross over-expression of CD30 that denotes neoplastic transformation. In normal physiology, a small population of activated T helper-2 (T_H-2) lymphocytes will express CD30, but this expression is at a much lower level. Mitogenic activation of lymphocytes *in vitro* cannot induce T lymphocytes to express the levels of CD30 present on MD lymphoma cells (Burgess and Davison, 2002).

CD30^{hi} cells occur in clusters in all MD lymphomas (Plate 3B), but are minority populations. The CD30^{hi} MD lymphoma cells are protected from cell death despite hyperproliferation, therefore fulfilling the accepted criteria for neoplastic transformation *in vivo*. These cells harbour the highest levels of MDV genome of all MD lymphoma cells (Burgess and Davison, 1999), yet they do not support productive infection of MDV. Critically, CD30^{hi} cells have large amounts of mRNA encoding the MDV putative oncogene Meq and a small RNA antisense to the cytolytic phase transcription factor, ICP4 (Ross et al., 1997). Furthermore, expression of Meq protein is directly proportional to CD30 expression; CD30^{hi} MD lymphoma cells have more Meq protein than any other cell in MD lymphomas.

The defined cell phenotype of MDV-transformed cells is CD30^{hi}, CD4$^+$, major histocompatibility complex (MHC) class I^{hi}, MHC class II^{hi}, interleukin-2 receptor α^+, CD28$^{lo/-}$ and T-cell-receptor $\alpha\beta^+$ (Burgess and Davison, 2002). This phenotype is typical of an activated memory T_H-2 lymphocyte. Notably, CD30^{hi} expression occurs on peripheral blood leukocytes (PBL) and lymphocytes infiltrating tissues at 5–7 dpi in both MD-resistant and MD-susceptible chickens. The proportions of these CD30^{hi} PBL and infiltrating lymphocytes does not significantly differ between MD-resistant and MD-susceptible chickens (Burgess et al., 2001; Burgess and Davison, 2002), suggesting that neoplastic transformation is an early event that occurs regardless of resistance to gross lymphomagenesis. This is not a new suggestion (see above). In MD-resistant chickens, the CD30^{hi} lymphocytes present in tissues die by apoptosis at the same time as CD8$^+$ lymphocytes become the predominant cell in these regressing lesions (Burgess et al., 2001).

Analysis of TCRβ chain gene family expression suggests that MD lymphomas have polyclonal origins (Burgess and Davison, 2002). This conflicts with the suggestion of MD clonality (Delecluse et al., 1993).

Anti-lymphoma immunity

There is no doubt that immunity can target MDV productively infected lymphocytes (see Chapter 10). In contrast, immunity to the neoplastically transformed lymphocytes in MD is poorly understood. However, much circumstantial evidence suggests that immunity to neoplastically transformed lymphocytes exists. The 'two-step hypothesis of MD immunity', i.e. immunity against MDV and immunity against the neoplastically transformed cells (Payne et al., 1976), has yet to be disproved. This topic was recently reviewed by Burgess and Venugopal (2002).

MD lymphoma regression

Although the work summarized in this section identifies lymphoma regression as a real phenomenon in MD and suggests a role for the immune system, none of the work definitively identifies an anti-lymphoma immune response or target antigens.

Nerve lesions progress from type A to B to C the longer individual chickens survive after the onset of MD-specific clinical signs (see Chapter 7 for details). A number of chickens that do not develop clinical signs, or gross lesions, have type C lesions. A shift from 'proliferative' to 'non-proliferative' histopathology corresponds with the onset of regression (Payne and Biggs, 1967; Lawn and Payne, 1979; Burgess et al., 2001). Lesion regression plays a role in the age-related resistance to MD (Sharma et al., 1973, 1975). This regression is T-cell dependent (Sharma et al., 1970, 1975) but independent of B cells, maternally-derived antibodies (Sharma and Witter, 1975) and neutralizing antibodies (Sharma, 1974).

Lymphoid infiltration into non-lymphoid organs has been quantified and related to clinical disease, and the cells involved in the lesions identified using immunohistochemistry (Burgess et al., 2001). All genetically susceptible line 7_2 chickens developed progressive MD. The moderately susceptible line 15_I chickens had fluctuating MD-specific clinical signs, with some individuals recovering (similar to the CNS pathology of persistent neurological disease described by Gimeno et al. (1999, 2001)). The resistant line 6_1 had a similar distribution of lymphoid infiltration into organs during the cytolytic phase (Plate 3C), although the number of cells involved was less than in line 7_2. When gross lymphomas were visible in line 7_2, histological lesions in line 6_1 were regressing. CD30^{hi} cells were present in lesions of resistant and susceptible genotypes from the cytolytic phase (Plate 3D); after this time their numbers increased in susceptible lines but decreased in resistant lines. In all chicken genotypes, in lesions with many CD8$^+$ lymphocytes AV37 immunostaining was weak (Plates 3E, 3F) and those AV37$^+$ cells were apoptotic. This suggests either that expression of the antigen was controlled by the CD8$^+$ cells, or that the AV37$^+$ cells themselves were.

Anti-MD lymphoma immunity

It is critical to emphasize that anti-tumour immunity is immunity directed against neoplastically transformed cells. Anti-tumour immunity is distinct from defining the immunogens themselves. In MD, the tumour antigens may theoretically be MDV proteins or host-encoded antigens.

Transplantable tumours

MDV-productive transplantable lymphoma systems have allowed observation of a superficial lymphoma without an initial systemic MDV infection (although chickens later become MDV infected). After inoculation, in syngeneic systems, there was initial growth of the transplant followed by regression at 14–18 dpi. Functioning T cells were necessary for this. Calnek *et al.* (1978) suggested the humoral immune system aided the response as administration of serum from infected convalescent chickens increased the rate of regression. Chickens vaccinated with transplantable lymphomas, which had lost their tumorigenic ability, were protected from lymphomas that would normally form after injection of the tumorigenic parental cell line (Nazerian and Witter, 1984). This was suggested to be an allo-response (Powell, 1983; Schat and Shek, 1984). In later similar work, though, syngeneic protection against a non-productive, highly virulent MDV transplantable lymphoma was demonstrated (Fletcher and Schierman, 1985; Tseng *et al.*, 1986; DiFronzo and Schierman, 1989).

Humoral immunity may also contribute to anti-lymphoma immunity (Calnek *et al.*, 1978). Pradhan *et al.* (1991) eluted antibodies from MD lymphomas, which recognized cultured *ex vivo*-derived lymphoma cells but not MDV antigens in the feather follicles. In addition, an ill-defined 'anti-MATSA antibody' blocked the immunohistochemical staining by the eluted antibody and *vice versa*. The authors suggested such antibodies could either enhance immunity by antibody dependent cell-mediated cytotoxicity (ADCC) or, conversely, aid immuno-evasion. Notably, ADCC against MDCC has never been demonstrated using serum from infected chickens, but has been demonstrated using serum from chickens immunized with lymphoblastoid cells (Powell, 1976). Dandapat *et al.* (1994) raised rabbit polyclonal antisera to MD-lymphoma cell-surface antigens that were 'most protective in vaccination experiments'. Goat polyclonal anti-idiotype antibodies were then produced against the original polyclonal rabbit antibodies. Also, rabbit polyclonal anti-idiotype antibodies were raised against the mouse RPH-6 mAb, which recognizes one (as yet undefined) MATSA. Anti-idiotype antibodies are specific to the 'idiotype' of the antibodies used to produce them, which means that the antibodies should mimic the 'shape' of the lymphoma antigens. Chickens vaccinated with either of these anti-idiotype antibodies were significantly protected against MD. The identity of the immunogen(s) is not known. Such immunity also has been demonstrated in human cancer systems (Spendlove *et al.*, 2000), and the mechanisms could be humoral or cell-mediated (Spendlove *et al.*, 2000; Durrant *et al.*, 2001).

Tumour antigens in MD

If anti-lymphoma immunity in MD exists, it could theoretically be directed against MDV antigens in the neoplastically transformed cells or host-encoded tumour antigens. One syngeneic cytotoxic lymphocyte (CTL) assay system, using transfected reticuloendotheliosis virus-transformed lymphocytes as targets, has been developed (Pratt *et al.*, 1992; Schat *et al.*, 1992). Specific CTL responses to some of the 80-plus predicted MDV antigens have been described (reviewed by Schat, 2001); however, most of these MDV antigens would not be present in MD lymphoma cells (Morgan *et al.*, 2001). MHC class I antigen expression is critical for CTL immunity. MHC class I antigen down-regulation occurs in productively-infected cells (Hunt *et al.*, 2001), and is a potential means of immune escape. In contrast, MHC class I antigen expression is higher in the neoplastically transformed MD-lymphoma cells than in any other lymphoma cells (Burgess and Davison, 2002). The neoplastically transformed cells, as far as is known, have the capacity to present peptide antigens to CTL.

The MDV antigens expressed in the neoplastically transformed MD lymphoma cells are yet to be defined (Morgan *et al.*, 2001). The MDV antigens so far investigated in CTL assays may not represent all those that elicit natural CTL responses. Any MDV-antigen translated in neoplastically transformed cells is potentially immunogenic.

Meq, the strongest candidate MDV oncogene (see Kung *et al.*, 2001; Chapter 4), is present at high levels in the neoplastically transformed MD lymphoma cells. Specific CTL responses to Meq have been demonstrated for both the MD-susceptible genotype $B^{19}B^{19}$ and the MD-resistant genotype $B^{21}B^{21}$. This work raises an obvious question: if CTLs exist, how do neoplastically-transformed cells survive in MD susceptible $B^{19}B^{19}$ chickens *in vivo*? Immune escape mechanisms must be involved in MD-susceptible chickens. In contrast, the putative MDV oncogene ORF-A (Kung *et al.*, 1995) produced no measurable CTLs in either $B^{21}B^{21}$ or $B^{19}B^{19}$ chickens (Omar and Schat, 1996). If ORF-A is oncongenic, it raises the opposite question: why don't lymphomas escape immunity in $B^{21}B^{21}$ chickens?

ICP4 has an interesting association between the presence of CTLs and host genotype. Only MD-resistant $B^{21}B^{21}$, but not MD susceptible $B^{19}B^{19}$, CTLs specifically lyse target cells *in vitro* (Omar and Schat, 1996). Although ICP4 is translated in lytically-infected CEF, the protein has not been demonstrated in latently infected cells. However, ICP4 mRNA is abundant in latently infected cells and may be translated (Morgan *et al.*, 2001). If it really is translated in neoplastically transformed MD lymphoma cells, the genetic difference in CTL responses may be highly significant.

Whether or not immunity against host-encoded tumour antigens exists in MD is controversial. Although auto-antibodies to MD lymphoma cells have been described (see above), the antigens have never been definitively identified as being host-encoded. Recently, the first host-encoded immunogen in MD lymphomas was identified as CD30 – i.e. natural humoral immunity to the over-expressed CD30 antigen occurs. Antibodies were demonstrated only in MHC B^2B^2 line 6_2 MD-resistant chickens that had been hyper-immunized with the vMDV strain HPRS-16. Even though anti-CD30 antibodies were detected, the actual immune response could be mediated by CTLs (see review by Chen, 2000). Regardless, prophylactic vaccines containing the CD30 gene delayed lymphomageneis in outbred chickens. The protective mechanism is not known.

Model for MD lymphomagenesis

Herpesviruses are masters of lifelong survival inside their hosts, and it is self-defeating for any pathogen to kill its host before the progeny can infect new hosts (Ewald, 1993). Host death is even less desirable for herpesviruses, whose strategy is lifelong persistence to achieve multiple progeny 'generations'. The most logical explanation for MDV-induced neoplastic transformation of lymphocytes in the first place is persistence in the host. However, gross lymphomas, at least if they grow so rapidly as to cause organ dysfunction, cannot be beneficial to MDV. Gross lymphomas must be incidental consequences of increasing the state of activation, lifespan and proliferative potential of latently infected lymphocytes.

Deregulated cell proliferation and protection from cell death define neoplastic transformation *in vivo* (Evan, 1997). Memory lymphocytes are the longest lived and most proliferative of all lymphocytes, and would therefore provide the longest persistence for lymphotrophic viruses. It is a small conceptual step to further increase proliferation and/or decrease death of memory lymphocytes, such that accumulation of these cells results (i.e. lymphoma). In support of this suggestion, CD30 signalling in normal lymphocytes promotes not only survival and proliferation but also cell death. However, CD30 expression in neoplastic cells promotes only survival and proliferation (Chiarle *et al.*, 1999; Tarkowski, 1999), and its up-regulation coincides with memory lymphocyte marker expression by T-helper 2 lymphocytes (Chiarle *et al.*, 1999; Tarkowski, 1999). MD lymphomas are hyperproliferative, antigen non-specific, non-clonal, 'super-memory' T-helper-2 lymphocytes, which are resistant to normal physiological cell death. This hypothesis has been developed from data presented above and also inferences from comparative herpesvirology.

A cellular model for Marek's disease lymphomagenesis is shown in Plate 4.

Summary

So far, the evidence shows that similar numbers of neoplastically transformed cells occur in both MD-resistant and MD-susceptible chickens soon after MDV infection. Somehow, the neoplastically transformed cells die in MD-resistant chickens but survive and proliferate in MD-susceptible chickens. This survival is despite expression of the largest amounts of the MDV Meq protein and high MHC class I and II expression (Burgess and Davison, 2002). CTLs recognizing Meq are present in both MD-susceptible and MD-resistant chickens. It is clear that the lymphoma environment must be critical for survival of the neoplastically transformed MD lymphoma cells in susceptible chickens (Burgess and Davison, 2002). CD30 is grossly over-expressed by the neoplastically transformed MD lymphoma cells. CD30$^+$ T helper lymphocytes normally secrete T-helper-2 cytokines, which antagonize cytotoxic lymphocytes (CTL). It is logical to hypothesize that the lymphoma environment is TH-2 biased, but this has never been demonstrated. Certainly, MD lymphoma cells could have other immune evasion strategies (e.g. CD28 down-regulation). For the first time, natural immunity has been demonstrated to an identified MATSA.

Regression of MD lymphomas is a documented phenomenon, and this regression probably has specific immune components. However, work is needed to confirm such immunity. Specifically, it is essential to identify the immunogens involved and understand how neoplastically-transformed MD lymphoma cells escape immunity. Harnessing anti-MD lymphoma immunity is an issue that unites vaccine technology (Chapter 13) and chicken genetics (Chapter 9). Controlling lymphoma growth (including CNS lesions) is a multifactorial problem.

Classical pathology and histopathology, polyclonal and monoclonal antibody technology and molecular biology all have their uses. However, with the sequencing and annotation of MDV genomes (Lee *et al.*, 2000; Tulman *et al.*, 2000) and the sequencing of the chicken genome, poultry research and production should enter the post-genomic era. Functional genomic technologies are becoming practicable for use in poultry research, together with the ability to identify neoplastically transformed cells in MD lymphomas. Understanding MD pathogenesis and MDV evolution towards greater virulence is now more achievable than ever before.

Acknowledgements

I would like to acknowledge the contributions by Drs N. Bumstead, G. Evan, T. F. Davison, J. Kaufman, H. -J. Kung, N. Osterrieder, M. S. Parcells, L. N. Payne, L. J. N. Ross, K. A. Schat, J. R. Young, R. L. Witter. Thanks also go to Dr J. K. L. Burgess for editorial input.

References

Akiyama, Y., Naito, M. and Kato, S. (1973). *Biken J.*, **16**, 91–94.
Barrow, A. D., Burgess, S. C., Howes, K. and Nair, V. K. (2003). *Avian Pathol.*, **32**, 183–191.
Benton, W. J. and Cover, M. S. (1957). *Avian Dis.*, **1**, 320–327.
Biggs, P. M., Purchase, H. G., Bee, B. R. and Dalton, P. J. (1965). *Vet. Rec.*, **77**, 1339–1340.
Bloom, S. B. (1970). *Avian Dis.*, **14**, 478–490.
Burgess, S. C. and Davison, T. F. (2002). *J. Virol.*, **76**, 7276–7292.
Burgess, S. C. and Venugopal, K. N. (2002). In *Advances in Medical and Veterinary Virology, Immunology and Epidemiology. Modern Concepts of Immunology in Veterinary Medicine: Poultry Immunology* (ed. T. Mathew), pp. 236–291. Thajema Publishing, West Orange, NJ.
Burgess, S. C. and Davison, T. F. (1999). *J. Virol. Methods*, **82**, 27–37.
Burgess, S. C., Basaran, B. H. and Davison, T. F. (2001). *Vet. Pathol.*, **38**, 129–142.
Burgess, S. C., Young, J. R., Baaten, B. J. G. *et al.* (2004). Submitted, for publication.
Calnek, B. W., Fabricant, J., Schat, K. A. and Murthy, K. K. (1978). *J. Natl Cancer Inst.*, **60**, 623–631.
Calnek, B. W., Carlisle, J. C., Fabricant, J. *et al.* (1979). *Am. J. Vet. Res.*, **40**, 541–548.
Calnek, B. W., Lucio, B., Schat, K. A. and Lillehoj, H. S. (1989). *Avian Dis.*, **33**, 291–302.
Chen, Y. T. (2000). *Cancer J.*, **6**, S208–S217.
Chiarle, R., Podda, A., Prolla, G. *et al.* (1999). *Clin. Immunol.*, **90**, 157–164.
Dandapat, S., Pradhan, H. K. and Mohanty, G. C. (1994). *Vet. Immunol. Immunopathol.*, **40**, 353–366.
Delecluse, H.-J. and Hammerschmidt, W. (1993). *J. Virol.*, **67**, 82–92.
Delecluse, H.-J., Schuller, S. and Hammerschmidt, W. (1993). *EMBO J.*, **12**, 3277–3286.
DiFronzo, N. L. and Schierman, L. W. (1989). *Intl J. Cancer*, **44**, 474–476.
Doak, R. L., Munnell, J. F. and Ragland, W. L. (1973). *Am. J. Vet. Res.*, **34**, 1063–1069.
Durrant, L. G., Parsons, T., Moss, R. *et al.* (2001). *Intl J. Cancer*, **92**, 414–420.
Evan, G. (1997). *Intl J. Cancer*, **71**, 709–711.
Ewald (1993). *Scientific American*, **268**, 86–93.
Fabricant, J., Calnek, B. W. and Schat, K. A. (1982). *Avian Dis.*, **26**, 257–264.
Fletcher, O. J. and Schierman, L. W. (1985). *Cancer Res.*, **45**, 1762–1765.
Gimeno, I. M., Witter, R. L. and Reed, W. M. (1999). *Avian Dis.*, **43**, 721–737.
Gimeno, I. M., Witter, R. L., Hunt, H. D. *et al.* (2001). *Vet. Pathol.*, **38**, 491–503.
Hoffmann-Fezer, G. and Hoffmann, R. (1980). *Vet. Immunol. Immunopathol.*, **1**, 113–123.
Hunt, H. D., Lupiani, B., Miller, M. M. *et al.* (2001). *Virology*, **282**, 198–205.
Kung, H. J., Tanaka, A. and Nonoyama, M. (1995). *Intl J. Oncol.*, **6**, 997–1002.
Kung, H. J., Xia, L., Brunovskis, P. *et al.* (2001). *Curr. Topics Microbiol. Immunol.*, **255**, 245–260.
Lampert, P., Garrett, R. and Powell, H. (1977). *Acta Neuropathol.*, **40**, 103–110.
Lawn, A. M. and Payne, L. N. (1979). *Neuropathol. Appl. Neurobiol.*, **5**, 485–497.

Lee, L. F., Wu, P., Sui, D. *et al.* (2000). *Proc. Natl Acad. Sci. USA*, **97**, 6091–6096.

Liotta, L. and Petricoin, E. (2000). *Nature Rev. Genet.*, **1**, 48–56.

Marek, J. (1907). *Dtsch Tierarztl. Wochenschr.*, **15**, 417–421.

McColl, K., Calnek, B. W., Harris, W. V. *et al.* (1987). *J. Natl Cancer Inst.*, **79**, 991–1000.

Moore, F. R., Schat, K. A., Hutchison, N. *et al.* (1993). *Intl J. Cancer*, **54**, 685–692.

Morgan, R. W., Xie, Q., Cantello, J. L. *et al.* (2001). *Curr. Topics Microbiol. Immunol.*, **255**, 223–224.

Murthy, K. K. and Calnek, B. W. (1979). *Avian Dis.*, **23**, 831–837.

Murthy, K. K., Dietert, R. R. and Calnek, B. W. (1979). *Intl J. Cancer*, **24**, 349–354.

Nazerian, K. and Witter, R. L. (1984). *Avian Dis.*, **28**, 160–167.

Omar, A. R. and Schat, K. A. (1996). *Virology*, **222**, 87–99.

Pappenheimer, A. M., Dunn, L. C. and Cone, V. (1926). *Bull. 143 Storrs Agricultural Exp. Stn*, 187–289.

Pappenheimer, A. M., Dunn, L. C. and Cone, V. (1929). *Exp. Med.*, **49**, 63–86.

Parcells, M. S. and Burgess, S. C. (2004). In *Comparative Aspects of Tumor Development*, Vol. 5 (ed. H. E. Kaiser). Kluwer Academic Publishers, Dordrecht (in press).

Payne, L. N. (1985). In *Marek's Disease* (ed. L. N. Payne), pp. 43–75. Martinus Nijhoff, Boston.

Payne, L. N. and Biggs, P. M. (1967). *J. Natl Cancer Inst.*, **39**, 281–302.

Payne, L. N. and Rennie, M. (1976). *Avian Path.*, **5**, 147–154.

Payne, L. N., Frazier, J. A. and Powell, P. C. (1976). *Int. Rev. Exp. Pathol.*, **16**, 59–154.

Powell, P. C. (1976). *Biblio. Haematol.*, **43**, 348–350.

Powell, P. C. (1983). *Avian Pathol.*, **12**, 461–468.

Powell, P. C., Payne, L. N., Frazier, J. A. and Rennie, M. (1974). *Nature*, **251**, 79–80.

Pradhan, H. K., Mohanty, G. C., Lee, W. Y. and Patnaik, B. (1991). *Vet. Immunol. Immunopathol.*, **29**, 229–238.

Pratt, J., Morgan, R. W. and Schat, K. A. (1992). *J. Virol.*, **66**, 7239–7244.

Prineas, J. W. and Wright, R. G. (1972). *Lab. Invest.*, **26**, 548–557.

Ross, L. J. N., Delorbe, W., Varmus, H. E. *et al.* (1981). *J. Gen. Virol.*, **57**, 285–296.

Ross, N., O'Sullivan, G., Rothwell, C. *et al.* (1997). *J. Gen. Virol.*, **78**, 2191–2198.

Schat, K. A. (2001). In *Current Progress on Marek's Disease Research* (eds K. A. Schat, R. M. Morgan, M. S. Parcells and J. L. Spencer), pp. 123–126. American Association of Avian Pathologists, Kennett Square, Pennsylvania.

Schat, K. A. and Calneck, B. W. (1978). *J. Natl Cancer Inst.*, **61**, 855–857.

Schat, K. A. and Shek, W. R. (1984). *Avian Pathol.*, **13**, 469–478.

Schat, K. A., Chen, C.-L. H., Calnek, B. W. and Char, D. (1991). *J. Virol.*, **65**, 1408–1413.

Schat, K. A., Pratt, W. D., Morgan, R. *et al.* (1992). *Avian Dis.*, **36**, 432–439.

Sharma, J. M. (1974). *Nature*, **247**, 117–118.

Sharma, J. M. (1981). *Am. J. Vet. Res.*, **42**, 483–486.

Sharma, J. M. (1985). In *Marek's Disease* (ed. L. N. Payne), pp. 43–75. Martinus Nijhoff, Boston.

Sharma, J. M. and Witter, R. L. (1975). *Cancer Res.*, **35**, 711–717.

Sharma, J. M., Davis, W. C. and Kenzy, S. G. (1970). *J. Natl Cancer Inst.*, **44**, 901–911.

Sharma, J. M., Witter, R. L. and Burmester, B. R. (1973). *Infect. Immun.*, **8**, 715–724.

Sharma, J. M., Witter, R. L. and Purchase, H. G. (1975). *Nature*, **253**, 477–479.

Spendlove, I., Li, L., Potter, V. *et al.* (2000). *Eur. J. Immunol.*, **30**, 2944–2953.

Tarkowski, M. (1999). *Arch. Immunol. Ther. Exp. (Warsz)*, **47**, 217–221.

Tseng, C. K., Fletcher, O. J. and Schierman, L. W. (1986). *Avian Pathol.*, **15**, 557–567.

Tulman, E. R., Afonso, C. L., Lu, Z. *et al.* (2000). *J. Virol.*, **74**, 7980–7988.

Wight, P. A. L. (1969). *J. Comp. Pathol.*, **79**, 563–570.

Witter, R. L., Stephens, E. A., Sharma, J. M. and Nazerian, K. (1975). *J. Immunol.*, **115**, 177–183.

Genetic resistance to Marek's disease

9

NAT BUMSTEAD* and JIM KAUFMAN
Institute for Animal Health, Compton Laboratory, Newbury, Berkshire, UK

Introduction

Differences in genetic resistance to Marek's disease (MD) were first reported by Asmundson and Biely (1932), who observed large differences in mortality among different families of chickens. These results were confirmed and extended by others, with the most extensive survey being carried out by Cole (1968), who compared susceptibility to the virulent (v) JM strain of MD among 28 commercial stocks of chickens and found mortality ranged from 18 per cent to 96 per cent among the lines tested. Hutt and Cole (1947) showed that it was possible to selectively breed lines for resistance or susceptibility to MD, even though at that time the causative agent was not known (see Chapter 2). The response to selection was very rapid, with mortality levels changing from 50 per cent in the initial population to 7 per cent and 94 per cent, respectively, in susceptible and resistant lines selected for only three generations (Cole, 1969). This very rapid response suggested resistance may be due to a small number of major genes. These experiments also confirmed that resistance was effective against a range of MD virus (MDV) isolates, in both experimental and natural infection. Since crosses between resistant and susceptible birds were intermediate in resistance, it seemed that allelic effects were likely to be additive.

The resistant and susceptible lines selected in these studies by Hutt and Cole at Cornell University (lines N and P) and Stone at East Lansing (lines 6 and 7; Stone, 1975) have largely provided the basis for investigating genetic resistance, although other studies have shown that similar differences exist in both outbred and commercial stocks.

The first progress towards understanding the genetic mechanisms responsible for resistance was made by Hansen *et al.* (1967), who observed an association between the inheritance of alleles at the *B* blood group locus and increased MD

---

*Nat Bumstead died on 20th April 2004 before publication of this book.

resistance in an outbred flock. Since the B blood group locus had been shown to be a marker for the chicken major histocompatibility complex (MHC), this suggested that genes within the MHC might be responsible. Hansen's results were subsequently confirmed (Longenecker *et al.*, 1976; Briles *et al.*, 1977) in experimental crosses between genetically typed chickens. Pazderka *et al.* (1975) were able to show that the difference in susceptibility of the selected N and P lines was largely associated with their MHC haplotypes.

MHC associated resistance

Many later studies confirmed that the B locus can have an enormous effect on the response to MDV infection (reviewed in Pazderka *et al.*, 1975; Longenecker and Mosmann, 1981; Calnek, 1985; Cole, 1985; Bacon, 1987; Schat, 1987; Plachy *et al.*, 1992; Bacon *et al.*, 2000). These studies vary widely in host genetics, sex, age and environment, as well as pathogen strain, dose and route of infection – all factors known to affect the level of host resistance. In most reports the contribution from the MHC can be separated from that of other loci, but in some reports the differences between lines are taken to indicate differences in MHC, without consideration of the many other background genes that are almost certainly involved. Given the many differences in genetics, environment and methodology, it is surprising that any consistent story could emerge. However, these many studies do, in fact, make a number of clear points.

The most important point is that there is a rough hierarchy of resistance determined by B haplotypes that is not restricted to the famous N and P lines described above (summarized in Briles *et al.*, 1977; Longenecker and Mosmann, 1981; Calnek, 1985; Bacon, 1987; Plachy *et al.*, 1992). Nearly all studies report that the B^{21} haplotype from whatever strain and genetic origin (white and brown leghorns, broiler chickens and red jungle fowl) confers strong resistance. Most studies report that the common haplotypes B^1, B^4/B^{13}, B^5, B^{12}, B^{15} and B^{19} (and others) are associated with susceptibility, with B^{19} generally (but not always) reported to confer the most susceptibility. The haplotypes B^2, B^6 and B^{14} generally are reported to confer moderate levels of resistance (Witter *et al.*, 1975; Longenecker and Mosmann, 1981; Calnek, 1985; Plachy *et al.*, 1992), although there seems to be wide variation. In most studies B^2 is associated with resistance nearly as good as or better than B^{21}. However, in some studies B^2 is at the low end of the susceptible haplotypes (Bacon and Witter, 1994), and in other studies B^{14} but not B^2 confers as much resistance as B^{21} (Longenecker *et al.*, 1976). The existence of this hierarchy of resistance determined by B haplotypes suggests that there are multiple alleles of at least one gene involved, rather than two alleles of a single gene.

In general, the resistant haplotypes are either dominant to, or co-dominant with, the susceptible haplotypes (Calnek, 1985), although there is a report of dominant susceptibility of a B^{19} haplotype (Hepkema *et al.*, 1993).

In several cases, the region responsible has been narrowed by recombination to the B-F/B-L region (Briles *et al.*, 1983; Plachy *et al.*, 1984; Hepkema *et al.*, 1993), which is known to contain the chicken MHC including the classical class I and class II B genes. A potential area of confusion is that another region encoding MHC-like genes, the Rfp-Y locus, has been reported in one study to confer resistance to MD. This single study examined the response to the very virulent (vv) MDV strain RB1B in the white leghorn population originally used to map the Rfp-Y locus, and found an approximately two-fold effect associated with one of the three Rfp-Y haplotypes (Wakenell *et al.*, 1996). Three other studies involving different mapping populations, and in some cases different virus strains, failed to find such an association (Vallejo *et al.*, 1997; Bumstead, 1998; Lakshmanan and Lamont, 1998).

Another major point is that the contribution of the B haplotype can be markedly affected by other factors, including host age, gender and other ('background') genes, as well as pathogen strain, dose and route of infection. For example, line 6 and line 7 (and their various sublines) both have the B^2 MHC haplotype, but vary enormously in MD susceptibility due to the effects of loci elsewhere in the genome (see later). Other examples include studies using three sets of MHC-congenic chicken lines (that is, chickens bred to have the same genome except for the MHC). Two haplotypes compared on the CC (Wellcome C line) background showed that B^{12} was more susceptible than B^4 (Plachy *et al.*, 1984); B^4 is now known to be equivalent to B^{13} (J. Kaufman, unpublished observations). Seven B haplotypes compared on the 003 background showed that B^2 was the most resistant, followed by B^{21} and B^Q (a B^{21} haplotype from red jungle fowl), and that B^{15}, B^{17}, B^{18}, B^{19} and B^{24} were susceptible when challenged with either v or vv MDV strains (Abplanalp *et al.*, 1985). Seven B haplotypes compared on the $15I_5$ background showed that B^{21} was the most resistant and that B^2, B^5, B^{12}, B^{13}, B^{15} and B^{19} were more susceptible when challenged with virulent strains, but that they were all susceptible when challenged with a vvMDV strain (Bacon and Witter, 1992). While these data with congenic lines are broadly consonant with the results from outbred lines, they demonstrate that the precise background of the host and the strain of virus can strongly influence the effect of the B haplotype. These results are not surprising (at least in retrospect). MDV has a relatively large genome and a complicated disease course, and so it would be a surprise if the host did not use a number of resistance strategies based on different gene systems. The MHC is a – perhaps at one time it was 'the' – major locus associated with resistance, and so it would be a surprise if MDV had not evolved, during the changes from classical to vv pathotypes, to escape or neutralize the effects of the MHC.

Structure and function of the MHC

In order to understand the genetics of MHC-determined resistance to MDV, it must be clear what is known about the structure and function of the chicken

MHC. Chicken chromosome 16, a microchromosome, contains the B locus, the Y (or Rfp-Y) locus and the nucleolar organizing region (NOR) (Bloom and Bacon, 1985; Briles *et al.*, 1993). The NOR contains many repeats of the ribosomal RNA genes, and is thought to be a region of high recombination located between the B and Y loci (Miller *et al.*, 1994; Fillon *et al.*, 1996). Indeed, the B and Y loci are unlinked in genetic crosses, and therefore are free to evolve independently, as though they were on different chromosomes. The Y locus contains apparently non-classical class I (Y-F) and class II B (Y-LB) genes, the numbers varying between haplotypes. The class Y-LB genes are non-polymorphic and are not expressed at a high level, while one Y-F gene has been shown to have several alleles with a non-classical peptide-binding site that is unlikely to bind pathogen-derived peptides (Zoorob *et al.*, 1993; Afanassieff *et al.*, 2001). The B locus has been divided into two parts by several well-publicized recombinants, with one part containing the polymorphic B-G genes found on red blood cells (along with most other B-G genes expressed on other cell types). The other part determines all of the functional attributes considered to be the definition of the MHC, including graft rejection, graft-versus-host reaction, mixed lymphocyte reaction, cellular collaboration etc. (Schierman and Nordskog, 1961; Pink *et al.*, 1977). A portion of this other part, the B-F/B-L region, has been sequenced and shown to be the chicken MHC (Kaufman *et al.*, 1999a, 1999b), containing the classical MHC class I (B-F) and class II B (B-LB) genes, among others (Figure 9.1).

Both functional and structural features of the chicken MHC have striking differences from what is known about mammals (Kaufman *et al.*, 1995, 1999a, 1999b; Kaufman, 1999). For instance, there are very strong disease associations with the human MHC but they involve autoimmune disease (or biochemical

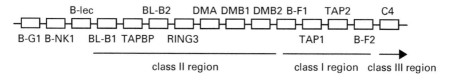

Figure 9.1 A portion of the sequenced region from the B locus. Open boxes represent genes with provisional gene names above the line (transcription from left to right) or below the line (transcription right to left). From left to right of figure: B-G1, the 8.5 gene; B-NK1 and B-lec, genes encoding type II membrane proteins with extracellular C-type lectin domains; B-LB1, gene encoding minor class II beta chain; TAPBP, gene encoding tapasin involving peptide loading to class I molecules; B-LB2, gene encoding major class II beta chain; RING3, gene encoding nuclear kinase; DMA, DMB1 and DMB2, genes encoding non-classical class II heterodimers involved in peptide loading to class II molecules; B-F1, gene encoding minor class I heavy chain; TAP1 and TAP2, genes encoding TAP heterodimer involved in pumping peptides from cytoplasm to endoplasmic reticulum; B-F2, gene encoding major class I heavy chain; C4, gene encoding serum complement component C4.

defects), whereas the associations with infectious pathogens (and with deliberate immunization such as vaccination) are relatively weak. In contrast, there are several examples of strong associations of chicken MHC haplotypes with resistance and susceptibility to infectious pathogens, and with response to live and inactivated vaccines. These functional differences are now explained, at least on some levels, by the structural differences between the MHC of a typical mammal and the chicken. Most importantly, many common haplotypes of the chicken MHC have a single dominantly expressed class I molecule and a single dominantly expressed class II molecule, the structural and functional characteristics of which limit the MHC-determined immune response. In contrast, most MHC haplotypes of a typical mammal express several class I and several class II molecules, so that there are more chances for an MHC-determined immune response than in a chicken.

This scenario has been best worked out for the T-cell response to pathogen- and vaccine-derived peptides bound to chicken class I molecules (Kaufman *et al.*, 1995, 1999a, 1999b; Kaufman, 1999). So, for instance, if the class I molecules expressed by an individual chicken fail to find, bind and present a protective peptide from a Rous sarcoma virus (RSV), then the chicken will not mount an MHC-determined response and will die from RSV-induced tumours. Similarly, an individual chicken will not be protected from a challenge of infectious bursal disease virus (IBDV) if the expressed class I molecules cannot present a peptide from a live sub-unit vaccine expressing the IBDV VP2 protein. The same lack of response would be true of a human individual whose MHC class I molecules failed to present a peptide from a particular pathogen or vaccine. However, there are three class I molecules expressed per human haplotype rather than one per chicken haplotype, with a correspondingly greater chance that a protective peptide will be bound and presented. (Indeed, the greater number of MHC molecules per human haplotype also increases the number of self-peptides that may be presented, and thus increases the chance of autoimmunity – explaining the strong associations between the human MHC and autoimmune diseases.)

The reason for this apparently suicidal strategy of chickens, in which individuals may die for want of an MHC molecule that finds, binds and presents a protective peptide to T cells, can be at least partially understood on the basis of structure (Kaufman *et al.*, 1995, 1999a, 1999b; Kaufman, 1999). The human MHC is broken up by frequent recombination, which makes sense given the vast size and large number of sequence repeats. In contrast, there is virtually no evidence for recombination within the chicken B-F/B-L region, which is small and simple. The evident lack of recombination over this region means that particular alleles of the MHC genes can stay together as haplotypes, co-evolving over significant periods of time to manifest certain phenotypes. For instance, the so-called transporter associated with processing (TAP) molecules pump potentially antigenic peptides from the cytoplasm to the lumen of the endoplasmic reticulum, where the peptides may bind to class I molecules.

In humans, the TAP genes and class I genes are located far apart and are separated by frequent recombination, with the consequence that the TAP molecules are not functionally polymorphic, pumping all peptides. In contrast, the TAP genes are next to the class I genes of chickens, with the consequence that the two can co-evolve. In fact, the chicken TAP genes are highly polymorphic, with the translocation specificity of individual TAP alleles converging with, or even controlling, the peptide motifs of the closely linked class I allele (B. Walker and J. Kaufman, unpublished observation). One consequence of this co-evolution is that the only effective class I molecules will be those with a peptide-binding specificity similar to the TAP translocation specificity, of which the most simple case will be a single dominantly expressed class I molecule. Interestingly, this seems to be a general situation among vertebrates other than mammals.

The single dominantly expressed class I molecule has other properties besides the specificity of peptides presented to T cells (Kaufman *et al.*, 1995). Unlike mammals, chicken MHC haplotypes determine striking differences in cell surface expression levels of class I molecules – as much as ten-fold on certain cell types. The reason is due to some aspect of transport of the dominantly expressed class I molecule from the inside of the cell to the surface. For all haplotypes examined, the dominantly expressed class I molecule is strongly expressed at the level of RNA and total protein, travelling to the surface very well in 'high expressing' haplotypes and less well in 'low expressing' haplotypes. For some 'low expressor' haplotypes, there is evidence for cell-surface expression of a second class I molecule (L. Hunt, F. Johnston and J. Kaufman, unpublished observations).

Mechanism of MHC-determined resistance to MD

One consequence of the low level of recombination within the B-F/B-L region is the difficulty of determining which gene(s) is responsible for MD resistance. A gene closely linked to the MHC that encodes a subunit of a trimeric G protein has been proposed (Guillemot *et al.*, 1989). Otherwise, the focus has been on the class I genes, which are highly polymorphic and could result in the hierarchy of responses seen in the infection studies cited above.

One group has shown that infection by all three MDV serotypes elicits low levels of cytolytic T lymphocytes (CTL) bearing CD8 and $\alpha\beta$ T cell receptors with specificity for a variety of MDV proteins, presumably as peptides bound to class I molecules. Most importantly, CTL with specificity for the immediate early protein ICP4 could be found in resistant B[21] chickens but not susceptible B[19] chickens, suggesting that a protective CTL response targets infected cells reactivating from latency (Omar and Schat, 1996, 1997; Markowski-Grimsrud and Schat, 2002). However, there is evidence from another group that reactivation of virus leads to rapid down-regulation of class I molecules on MDV tumour cells (Hunt *et al.*, 2001). Recently, Garcia-Camacho *et al.* (2003) reported that natural killer (NK) cells may be very important.

A third group has reported that, in contrast to mammals, the relative level of expression of chicken class I molecules at the surface of uninfected cells varies considerably (Kaufman *et al.*, 1995; Kaufman and Salomonsen, 1997). Remarkably, the rank order of cell surface expression reflects the consensus hierarchy of response by different MHC haplotypes, with the most susceptible B[19] haplotype having the highest expression, the most resistant B[21] haplotype having the lowest expression, and the others ranged in between. It remains unclear which mechanism of resistance is implied by the range of expression levels, with at least ten possibilities (Kaufman and Salomonsen, 1997). One strong possibility is that the effectors are NK cells, which in mammals detect differences in cell surface expression levels (Moretta *et al.*, 1992). Interestingly, the chicken MHC, unlike that of mammals, includes two genes with the general structure of lectin-like NK cell receptors, which might co-evolve with and recognize the class I molecules in the same haplotype (Kaufman *et al.*, 1999a, 1999b; S. Rogers and J. Kaufman, unpublished observations). One possibility is that both CTLs and NK cells recognize the class I molecules to confer protection from MDV infection, perhaps at different times in the course of the disease. Alternatively, some other gene in this region could be involved in resistance and susceptibility to MD.

Non-MHC associated resistance

Despite the strong contribution of some MHC haplotypes, it is clear that other genes also have a strong influence on the overall level of MD resistance. This is particularly evident when comparing the inbred lines 6 and 7 and their sublines, developed at East Lansing by Stone (1975). Although these lines are homozygous for the same MHC (B[2]) haplotype, they differ greatly in their resistance to a wide range of MDV strains (Pazderka *et al.*, 1975). Equally, in outbred and commercial flocks it has been apparent that the MHC associated element of resistance does not always fully explain the observed differences in resistance (Hartmann *et al.*, 1992). A number of approaches have been used to investigate the genetic mechanisms underlying the non-MHC associated aspect of resistance, which have largely focused on comparisons between lines 6 and 7, either at the phenotypic level, by investigation of candidate genes or, more recently, through genomic mapping.

Differences in viral replication between lines 6 and 7

The different resistance between lines 6 and 7 could be caused by differences in the replication or spread of MDV, differences in the likelihood of an infected cell becoming transformed, or differences in the immune control of either infection or developing tumours. Because of the highly cell-associated nature of MDV, early studies compared levels of virus replication using plaque assays, in which infected leukocytes were plated onto reporter cells. These studies showed levels of viral replication were considerably higher in line 7 from very early in infection (Lee *et al.*, 1981), though this result could be due either to greater numbers of infected cells or to a greater ability of these cells to initiate plaques. More recently a quantitative polymerase chain reaction (PCR) technique has been developed to measure levels of virus throughout infection (Bumstead *et al.*, 1997). This approach has simplified measurements of viral levels and confirmed the higher levels of replication of MDV in line 7, indicating that these are indeed due to greater numbers of infected cells. In crosses between lines 6 and 7, levels of virus present in lymphocytes early in infection correlate well with the later development of tumours. This suggests viral load is a critical aspect of the difference between these lines, although other genetic differences could also affect resistance by controlling tumour development.

Lymphocytes from line-7 chickens contained much larger numbers of viral copies than the equivalent line-6 chickens, from very early in infection (day 4) through to the time of death. While in line-7 chickens MD viraemia increased to a very high level and remained at this level, in line-6 chickens viral levels peaked at around 10 days and fell to below detectable levels thereafter. This very early difference suggests the differences between these two lines are of an innate nature rather than an adaptive response. The later clearance of infection in line 6 is more suggestive of an adaptive response, though whether the failure of line-7 chickens to clear infection is due to greater initial damage to the immune system (Baigent and Davison, 1999) or underlying differences in their capacity for mounting an immune response is unclear (Bumstead *et al.*, 1997).

Phenotypic comparisons

Initial comparisons between lines 6 and 7, in default of molecular markers, compared immunological responses between lines. These results showed that the difference between the two lines was not likely to be due to the ability of MDV to replicate within their cells, since levels of replication of the virus in fibroblast cultures made from the two lines were similar (Gallatin and Longenecker, 1979). Bursectomy or bursal transfers had little effect on resistance (Powell *et al.*, 1986), and comparisons showed increased phagocytic activity following MDV infection, that was initially greater in line 6 (Powell *et al.*, 1983). However using herpesvirus of turkeys (HVT), Gallatin and Longenecker (1979) were able to show that spleen cells from line-7 birds were able to adsorb more than six times more virus than corresponding cells from line 6. In addition, transplantation of thymus tissue from newly-hatched line-7 chicks into thymectomized line-6 chickens raised the level of susceptibility in the chimaeric recipients, whereas the reciprocal transfer of line-6 thymus tissue into line-7 chicks had no effect. Powell *et al.* (1982) showed that in reciprocal thymic transfers between line 6 and line 7, levels of leukocyte viraemia corresponded to the donor rather than the recipient, and correlated with susceptibility to the disease. Gallatin and Longenecker (1979) had originally proposed that this difference in susceptibility might be due to thymic education, since transfer of embryonic spleen cells without thymus did not affect susceptibility. However, Powell *et al.* (1982) were able to show, by transferring cells between animals of different sexes, that in line-6 recipients of line-7 thymuses, in the majority of cases the resulting tumours were of donor origin. This was the case even though the transferred thymuses rapidly became depleted of T cells, before becoming repopulated with cells of line-6 origin. These results suggest that susceptibility may be due, at least in part, to the greater susceptibility of line-7 lymphocytes to infection, leading to greater MDV proliferation, larger numbers of infected T cells and increased transformation.

Investigation of candidate genes

Several attempts have been made to address the differences between lines 6 and 7 by identifying lymphocyte surface markers which differ between the lines, since these might have a direct role in resistance by influencing lymphocyte function. Fredericksen *et al.* (1977) prepared allo-antisera by cross-immunizations using lymphocytes of the two lines and identified an alloantigen expressed on lymphocytes, which he designated Ly-4. Similarly, Gilmour *et al.* (1976) identified two further loci, Bu-1 and Th-1, by cross-immunizations of bursal cells and thymocytes, respectively, between the two lines. For each of these loci experiments suggested a degree of association with resistance, although in the absence of genomic maps it was not possible to determine if the loci themselves were responsible or if they were linked to resistance loci. The genes responsible for Ly-4 and Th-1 are as yet unknown; however, recently Tregaskes *et al.* (1996) have cloned a gene (chB6) which encodes Bu-1. chB6 has weak similarity in sequence and chromosomal location to the mammalian CD48 family of lymphocyte regulating surface receptors (C. A. Tregaskes and J. R. Young, IAH, personal communication). In chickens, chB6 lies in the central portion of chicken chromosome 1 and showed a weak association with resistance in crosses between line 6 and line 7, supporting the original observations of Gilmour *et al.* (1976). It remains uncertain whether this effect is due to the chB6 gene itself or to other linked genes, and at present it is not clear which other likely candidate genes might lie in this region. More recently, the increasing availability of identified chicken genes has made it possible to investigate a number of other candidate genes predicted to play a role in the response to infection, including interferons, cytokines and other immunoregulatory molecules. Among these the most striking association was that seen between polymorphisms in the growth hormone (GH) gene among 12 strains of white leghorns by Kuhnlein *et al.* (1997).

Genomic mapping approaches

With the development of a genomic map for the chicken it has become possible to locate and identify the genes responsible for resistance in a more general manner by genomic mapping and positional cloning. Vallejo *et al.* (1998) exploited this by carrying out a genetic mapping experiment in an F2 cross between the lines 6_3 and 7_2. Chickens were challenged with the vMDV strain, JM, at 1 week of age, and monitored for evidence of disease until 10 weeks of age, when they were examined *post mortem* for evidence of neural or visceral lesions. Resistance was scored in terms of a range of parameters, including viraemia and the numbers and types of organs found to be affected by either tumours or neural lesions at necropsy. A panel of microsatellite markers was used to geno-type the birds, and association compared between markers and the parameters of disease, both individually and in combination. From these comparisons 14 chro-mosomal regions were identified (Vallejo *et al.*, 1998; Yonash *et al.*, 1999) which showed association with one or more aspects of resistance, even though marker coverage of the genome was incomplete at 70 per cent. In general, resistance was dominant and, although the effects of individual regions were small in total, they accounted for 75 per cent of the variation between the parent lines.

Using levels of viral load and tumour development as measures of suscepti-bility, Bumstead (1998) attempted to map resistance in a line $(6_1 \times 7_2) \times 7_2$ backcross population which was infected with the vMDV strain, HPRS16. This population was investigated both using a panel of conventional microsatellite markers and by targeted comparisons using a representational difference analysis (RDA) approach (Wain *et al.*, 1998). To generate RDA differences, a resistant line-6 bird was compared with a pool of F2 birds identified as susceptible in terms of viral load and tumour development. Of the 8 RDA clones identified and mapped in this population, 4 lie in one region of chicken chromosome 1. This region shows significant levels of association with resistance in the mapped birds of the backcross population, in terms of both reduced viral levels and reduced tumour incidence. As the viral levels are measured early in infection, well before tumour development, this implies that the mode of action of the resistance locus is through control of viraemia development. This association seen in the back-cross mapping panel was confirmed in an independent F2 population, typed only for this locus, providing strong evidence that this region of chromosome 1 con-tains at least one resistance gene, which was designated MDV1.

Comparison of genes lying in this region in chickens with human and mouse indicated that it corresponds to regions of conserved synteny on human chromo-some 12 and mouse chromosome 6 (Plate 5). Strikingly, in mammalian species these regions contain the human and murine NK cell gene clusters, which con-tain multiple lectin-like cell surface proteins important in controlling the activity of NK cells (Yabe *et al.*, 1993, reviewed in Moretta *et al.*, 2001). In mice, sus-ceptibility to the murine herpesvirus cytomegalovirus (CMV) has been mapped to the NK complex (NKC) region (Scalzo *et al.*, 1990, 1995), and recently shown to be due to absence of the Klra8 gene within the NK lectin cluster (Brown *et al.*, 2001; Lee *et al.*, 2001). The phenotype of reduced viral replication in the spleens of CMV-infected animals is similar to that of MDV1 resistance. Since NK cells have long been thought to play a role in the early response to MD (Sharma, 1981, reviewed in Schat and Markowski-Grimsrud, 2001), genes affecting their activity would be likely candidate resistance genes. Recently, a chicken gene with similarity to the mammalian NKC-type lectin-like genes has been identified in this region (M. Clendenning and N. Bumstead, unpublished results). However, it has yet to be shown that this has similar functions to the mammalian genes, or that chickens have a wider NKC-regulating cluster corresponding to that of mammals. Other genes close to this region in mammals, such as CD4 and genes of the TNF receptor family, are also possible candidate genes.

In a complementary approach to determining the genes underlying these areas of association with resistance, Liu *et al.* (2001a) have used a DNA microarray to assess differences in transcription level of a panel of 1200 genes in uninfected and MDV-infected peripheral blood lymphocytes of lines 6_3 and 7_2. A number of genes were identified which showed two-fold or greater differences in expression between the two lines, including known immunologically relevant genes such as TCR-β, MHC class I, interferon γ and immunoglobulin light chain, but also many genes of currently unknown function. Relating the genomic positions of the differentially regulated genes to their genomic location identified cases where the genes coincided with regions associated with resistance to MD, notably in the case of lymphotactin on chromosome 1 in a resistance region identified by Yonash *et al.* (1999) and on the edge of that identified by Bumstead (1998).

In another novel approach to the identification of resistance genes, Liu *et al.* (2001b) have sought to utilize the interactions of a gene from MDV itself to identify host genes involved in the infection process. Recognizing that the viral SORF2 gene is inactivated in the non-oncogenic MDV strain RM1 (Jones *et al.*, 1996), and hence may be important in viral oncogenicity, Liu *et al.* (2001b) expressed the SORF2 gene to find the range of proteins interacting with it in a yeast 2-hybrid assay. The expressed product was used to screen a splenic cDNA library and identify those chicken genes that bind to the SORF2 protein (Tulman *et al.*, 2000). The assay identified chicken growth hormone (GH) as interacting with SORF2, and this was confirmed by co-immunoprecipitation and histological co-localization. Interestingly, chicken GH was among the genes identified as being differentially up-regulated in the microarray experiments of Liu *et al.* (2001a), and had previously been shown by Kuhnlein *et al.* (1997) to show distorted allele frequencies in lines of chickens selected for MD resistance. A polymorphism in the fourth intron of GH was used to genotype a chicken population infected with MDV, and an association was seen between alleles at the GH locus and the number of tumours and length of survival, although only when the population was partitioned into MHC classes.

Conclusions

It is clear that a number of genes including the chicken MHC make important contributions to MD resistance. An important question, and one that still remains unanswered, is how these different sets of genes interact to modulate MD resistance/susceptibility in the individual chicken. With present concerns over increasingly aggressive MD vaccination regimes driving MDV to evolve to increased levels of virulence, there is renewed interest using genetic selection to improve MD resistance of modern breeding stock. This can be approached by conventional selection and breeding; however, the availability of new genomic, transcriptomic and proteomic technologies – DNA arrays and interference RNA (RNAi), together with the availability of the chicken genome sequence – should revolutionize the identification of MD resistance genes. Chicken genetics may provide more sustainable means of preventing outbreaks of MD in the future (see Chapter 14).

Acknowledgements

We are grateful to Mark Clendenning for providing Plate 5, which shows details of the chicken NKC region.

References

Abplanalp, H., Schat, K. A. and Calnek, B. W. (1985) In *Proceedings of the International Symposium on Marek's Disease, Ithaca, NY* (eds B. W. Calnek and J. L. Spencer), pp. 347–358. American Association of Avian Pathologists, Kennett Square, Pennsylvania.

Afanassieff, M., Goto, R. M., Ha, J. *et al.* (2001). *J. Immunol.*, **166**, 3324–3333.

Asmundson, V. S. and Biely, J. (1932). *Can. J. Res.*, **6**, 171–176.

Bacon, L. D. (1987). *Poultry Sci.*, **66**, 802–811.

Bacon, L. D. and Witter, R. L. (1992). *Avian Dis.*, **36**, 378–385.

Bacon, L. D. and Witter, R. L. (1994). *Avian Dis.*, **38**, 65–71.

Bacon, L. D., Hunt, H. D. and Cheng, H. H. (2000). *Curr. Topics Microbiol. Immunol.*, **255**, 121–137.

Baigent, S. J. and Davison, T. F. (1999). *Avian Pathol.*, **28**, 287–300.

Bloom, S. and Bacon, L. (1985). *J. Heredity*, **76**, 146–154.

Briles, W. E., Stone, H. A. and Cole, R. K. (1977). *Science*, **195**, 193–195.

Briles, W., Briles, R., Taffs, R. and Stone, H. (1983). *Science*, **219**, 977–979.

Briles, W. E., Goto, R., Auffray, C. and Miller, M. (1993). *Immunogenetics*, **37**, 408–414.

Brown, M. G., Dokun, A. D., Heusal, J. W. *et al.* (2001). *Science*, **292**, 934–937.

Bumstead, N. (1998). *Avian Pathol.*, **27**, S78–S81.

Bumstead, N., Silibourne, J., Rennie, M. *et al.* (1997). *J. Virol. Methods*, **65**, 75–81.

Calnek, B. W. (1985). In *Marek's Disease* (ed. L. N. Payne), pp. 293–328. Martinus Nijhoff, Boston.

Cole, R. K. (1968). *Avian Dis.*, **12**, 9–28.

Cole, R. K. (1969). *NY Food Life Sci.*, **2**, 11–13.

Cole, R. K. (1985). In *Proceedings of the International Symposium on Marek's Disease, Ithaca, NY* (eds B. W. Calnek and J. L. Spencer), pp. 318–329. American Association of Avian Pathologists, Kennett Square, Pennsylvania.

Fillon, V., Zoorob, R., Yerle, M. *et al.* (1996). *Cytogenet. Cell Genet.*, **75**, 7–9.
Fredericksen, T. L., Longenecker, B. M., Pazderka, F. *et al.* (1977). *Immunogenetics*, **5**, 535–552.
Gallatin, W. M. and Longenecker, B. M. (1979). *Nature*, **280**, 587–589.
Garcia-Camacho, L., Schat, K. A., Brooks, R. and Bounous, D. I. (2003). *Vet. Immunol. Immunopathol.*, **15**, 145–153.
Gilmour, D. G., Brand, A., Donnelly, N. and Stone, H. S. (1976). *Immunogenetics*, **3**, 549–563.
Guillemot, F., Billault, A. and Auffray, C. (1989). *Proc. Natl Acad. Sci.* USA., **86**, 4594–4598.
Hansen, M. P., van Zandt, J. N. and Law, G. R. J. (1967). *Poultry Sci.*, **46**, 1268.
Hartmann, W. (1988). *J. Anim. Breed.Genet.*, **105**, 709–716.
Hartmann, W., Hala, K. and Heil, G. (1992). *Arch. Tierzucht*, **35**, 169–180.
Hepkema, B. G., Blankert, J. J., Albers, G. A. A. *et al.* (1993). *Anim. Genet.*, **24**, 283–287.
Hunt, H. D., Lupiani, B., Miller, M. M. *et al.* (2001). *Virology*, **282**, 198–205.
Hutt, F. B. and Cole, R. K. (1947). *Science*, **106**, 379–384.
Jones, D., Brunoushis, P., Witter, R. and Kung, H. J. (1996). *J. Virol.*, **700**, 2460–2467.
Kaufman, J. (1999). *Immunogenetics*, **50**, 228–236.
Kaufman, J. and Salomonsen, J. (1997). *Hereditas*, **127**, 67–73.
Kaufman, J., Völk, H. and Wallny, H. J. (1995). *Immunol. Rev.*, **143**, 63–88.
Kaufman, J., Jacob, J., Shaw, I. *et al.* (1999a). *Immunol. Rev.*, **167**, 101–118.
Kaufman, J., Milne, S., Goebel, T. W. F. *et al.* (1999b). *Nature*, **401**, 923–925.
Kuhnlein, U., Ni, L., Weigend, S., Gavora, J. S. *et al.* (1997). *Anim. Genet.*, **28**, 116–123.
Lakshmanan, N. and Lamont, S. J. (1998). *Poultry Sci.*, **77**, 538–541.
Lee, L. F., Powell, P. C., Rennie, M. *et al.* (1981). *J. Natl Cancer Inst.*, **66**, 789–796.
Lee, S. H., Girard, S., Macina, D. *et al.* (2001). *Natl Genet.*, **28**, 42–45.
Liu, H. C., Cheng, H. H., Tirunagaru, V. *et al.* (2001a). *Anim. Genet.*, **32**, 351–359.
Liu, H. C., Kung, H. J., Fullen, J. E. *et al.* (2001b). *Proc. Natl Acad. Sci. USA*, **98**, 9203–9208.
Longenecker, B. M. and Mosmann, T. R. (1981). *Immunogenetics*, **13**, 1–23.
Longenecker, B. M., Pazderka, F., Gavora, J. S. *et al.* (1976). *Immunogenetics*, **3**, 401–407.
Markowski-Grimsrud, C. J. and Schat, K. A. (2002). *Vet. Immunol. Immunopathol.*, **90**, 133–144.
Miller, M. M., Goto, R., Taylor, R. L. *et al.* (1994). *Proc. Natl Acad. Sci. USA*, **93**, 3958–3962.
Moretta, A., Bottino, C., Vitale, M. *et al.* (2001). *Annu. Rev. Immunol.*, **19**, 197–223.
Moretta, L., Ciccone, E., Moretta, A. *et al.* (1992). *Immunol. Today*, **13**, 300–306.
Omar, A. R. and Schat, K. A. (1996). *Virology*, **222**, 87–99.
Omar, A. R. and Schat, K. A. (1997). *Immunology*, **90**, 579–585.
Pazderka, F., Longenecker, B., Law, G. and Ruth, R. (1975). *Immunogenetics*, **2**, 101–130.
Pink, J. R. L., Droege, W., Hala, K. *et al.* (1977). *Immunogenetics*, **5**, 203–216.
Plachy, J., Jarajda, V. and Benda, V. (1984). *Folia Biol. (Praha)*, **30**, 251–258.
Plachy, J., Pink, J. and Hala, K. (1992). *Crit. Rev. Immunol.*, **12**, 47–79.
Powell, P. C., Lee, L. F., Mustill, B. M. and Rennie, M. (1982). *Intl J. Cancer*, **29**, 157–174.
Powell, P. C., Hartley, K. J., Mustill, B. M. and Rennie, M. (1983). *J. Reticuloendothelial Soc.*, **34**, 289–297.
Powell, P. C., Irving, N. G., Prynne, A. P. and Rennie, M. (1986). *Avian Pathol.*, **15**, 597–609.
Scalzo, A. A., Fitzgerald, N. A., Simmons, A. *et al.* (1990). *J. Exp. Med.*, **171**, 1469–1483.
Scalzo, A. A., Lyons, P. A., Fitzgerald, N. A. *et al.* (1995). *Genomics*, **27**, 435–441.

Schat, K. A. (1987). In *Avian Immunology: Basis and Practice*, Vol. 2 (ed. A. Toivanen and P. Toivanen), pp. 101–128. CRC Press, Boca Raton.

Schat, K. A. and Markowshi-Grimsrud, C. J. (2001). *Curr. Topics Microbiol. Immunol.*, **255**, 91–120.

Schierman, L. W. and Nordskog, A.W. (1961). *Science*, **134**, 1008–1009.

Sharma, J. M. (1981). *Avian Dis.*, **25**, 882–893.

Stone, H. A. (1975). *USDA Tech. Bull. No. 1514*, United States Department of Agriculture, Washington DC.

Tregaskes, C. A., Bumstead, N., Davison, T. F. and Young, J. R. (1996). *Immunogenetics*, **44**, 212–217.

Tulman, E. R., Afonso, L. L., Lu, Z. *et al.* (2000). *J. Virol.*, **74**, 7980–7988.

Vallejo, R. L., Bacon, L. D., Liu, H. C. *et al.* (1988). *Genetics*, **48**, 349–360.

Vallejo, R. L., Pharr, G. T., Liu, H. C. *et al.* (1997). *Anim. Genet.*, **28**, 331–337.

Wain, H. M., Toye, A. A., Hughes, S. and Bumstead, N. (1998). *Anim. Genet.*, **29**, 446–452.

Wakenell, P. S., Miller, M. M., Goto, R. M. *et al.* (1996). *Immunogenetics*, **44**, 242–245.

Witter, R. L., Stephens, E. A., Sharma, J. M. and Nazerian, K. (1975). *J. Immunol.*, **115**, 177–183.

Yabe, T., McSherrry, C., Bach, F. H. *et al.* (1993). *Immunogenetics*, **37**, 455–460.

Yonash, N., Bacon, L. D., Witter, R. L. and Cheng, H. (1999). *Anim. Genet.*, **30**, 126–135.

Zoorob, R., Bernot, G., Renoir, D. M. *et al.* (1993). *Eur. J. Immunol.*, **23**, 1139–1145.

Immunity to Marek's disease

10

FRED DAVISON and PETE KAISER

Institute for Animal Health, Compton Laboratory, Newbury, Berkshire, UK

Introduction

Marek's disease herpesvirus (MDV) targets a number of different cell types during its life cycle. Lymphocytes play an essential role, although within them virus production is restricted and only virions, not fully enveloped viruses, are produced (semi-productive infection, see Chapter 6). MDV is highly cell-associated, so its genome can only be passed on to daughter lymphocytes at cell division or transmitted to other cells by direct contact (Kaleta and Neumann, 1977). Targeting lymphocytes is insidious, for these are, or become, the effector cells of the immune system. Hence in MDV-infected birds, some lymphocytes contribute to protective immune responses against MDV whilst others are involved in spreading the virus and causing pathology. By debilitating the immune system and causing immunosuppression (see Chapter 11), MDV makes its host more vulnerable to tumour development as well as to other pathogens. All chickens are susceptible to MDV infection, and vaccination is essential to protect the susceptible host from developing clinical disease. Nevertheless, MDV infects and replicates in vaccinated chickens, with the challenge virus being shed from the feather-follicle epithelium.

Details of the transmission and life cycle of MDV are fully described in Chapter 6. However, it is pertinent to point out a few essential details relevant to the immune response. Once MDV has been taken up by phagocytic cells, it becomes strictly intracellular and is transported to the lymphoid tissues where it causes a cytolytic infection first in B lymphocytes and then activated T lymphocytes. Overall this cytolytic phase lasts for about 7 days, and is most likely curtailed by the development of adaptive immune responses. MDV then enters latency predominantly in CD4$^+$ T lymphocytes, though other lymphocyte types may also be latently infected (Calnek *et al.*, 1984). Infected lymphocytes carry MDV to the feather-follicle epithelium, where a fully productive infection occurs with infectious cell-free MDV being shed with flakes of skin from about 10 days post-infection (dpi) until the end of the host's life. In resistant genotypes MDV latency in lymphocytes is maintained for life, with reactivation occurring in some cells from time to time. Immunosurveillance is important for maintaining latency, and treatment with immunosuppressive drugs causes MDV to become reactivated (Buscaglia *et al.*, 1988) and breaks vaccinal protection (Powell and Davison, 1986). In the susceptible host MDV can transform CD4$^+$ T lymphocytes, causing lymphomas to form in various visceral tissues and/or neural lesions that result in paralysis. The outcome of infection with MDV therefore depends on a complex interplay of factors involving the MDV pathotype and the host genotype. Host factors that influence the course of MD are predominantly the responses of the innate and acquired immune systems, and these are modulated by: age at infection and maturity of the immune system; vaccination status; co-infection with other (immunosuppressive) pathogens; the sex of the host; and various physiological factors (see Chapter 6).

MDV entry via the lungs

Chickens become infected with MDV by inhaling infected dust and dander shed in the form of flakes of skin from MDV-infected birds. The lung is almost certainly the portal for virus entry (see Chapter 6). However, almost all studies on immune responses to MDV have involved inoculating the virus by a parenteral route, thus ignoring the very early events following natural infection. MDV-infected dust insufflated into the lower trachea causes MDV infection (unpublished observations) though the actual site(s) of uptake along the airways, and the cellular mechanisms involved in virus uptake, have yet to be identified (see Chapter 6). Moreover, the avian lung has a very different structure from its mammalian counterpart. Being an open system it lacks alveoli, and gas exchange takes place in capillaries that form a mantle around the tertiary bronchi (parabronchi). Due to the narrowness of these capillaries there are few airway-resident macrophages (Toth, 2000), so MDV must somehow cross the lung epithelial lining before being transported by phagocytic cells to the lymphoid tissues. Almost nothing is known about innate or adaptive mucosal immune responses in the avian pulmonary system. Identification of soluble factors such as interferons and immunoglobulins, other molecules such as defensins as well as pulmonary macrophages, dendritic cells and lymphocytes and their receptors, is essential for understanding the early events after MDV entry and for devising new strategies, or novel vaccines, that interfere with MDV uptake.

Role of macrophages

Macrophages have a number of essential roles in MDV infection. They are carrier cells that phagocytose MDV in the respiratory tract and transport it to the lymphoid tissues, where it replicates in lymphocytes (Barrow *et al.*, 2003). *In vitro* studies have indicated that macrophages, unlike lymphocytes, are refractory to MDV infection (Haffer *et al.*, 1979). However, recently it was reported that during the cytolytic phase splenic macrophages can express MDV antigens consistent with the virus replicating in these cells (Barrow *et al.*, 2003).

Macrophages play a pivotal role in innate immunity and the development of adaptive immune responses by acting as antigen-presenting cells. Innate immune responses are rapidly induced by viral infection, controlling early events until acquired immune responses become fully effective several days later. Macrophages recognize antigen through pattern recognition receptors and release a variety of cytokines (see later) as well as soluble factors such as nitric oxide (NO). Recent evidence, from both *in vitro* and *in vivo* studies, indicates that NO produced through the increased activity of inducible nitric oxide synthase (iNOS) is important for inhibiting MDV replication in the cytolytic phase of infection and continues well into the latent phase (Xing and Schat, 2000a; Djeraba *et al.*, 2002a). Earlier and increased levels of iNOS expression and NO production occur in MD-resistant genotypes, and probably contribute to the reduced MD viraemia and lower levels of clinical disease associated with genetic resistance (see Chapter 9).

Macrophages obtained from MDV-infected chickens inhibit MDV replication *in vitro* and are much more effective than macrophages obtained from uninfected birds (Kodama *et al.*, 1979), whereas depleting macrophages from MDV-infected spleen cell preparations enhances the amount of virus that can be isolated by co-culture. After MDV infection macrophage activity becomes increased (Powell *et al.*, 1983), with higher levels of phagocytic activity obtained from genetically resistant than from susceptible chickens. Treatments that decrease macrophage numbers or activity, such as anti-macrophage serum, multiple injections of silica, levamisole or dichloromethylene biphosphonate encapsulated in liposomes, increase the incidence of MD tumours and tumour load (Higgins and Calnek, 1976; Haffer *et al.*, 1979; Djeraba *et al.*, 2002a). Treatments that increase macrophage numbers, such as a single injection of silica, or brewer's thioglycollate broth, which increases macrophage activity, decreased the incidence of MD tumours (Gupta *et al.*, 1989).

It is clear that macrophages play an important role in reducing MD viraemia in the early cytolytic phase, hence reducing the likelihood of clinical MD developing, but their role in the tumour phase is less clear. Sharma (1983) isolated an adherent cell population with the characteristics of macrophages from MD tumours, and these were able to kill MDV-transformed chicken cell-line (MDCC) cells in an 18-h chromium release assay. The presence of macrophages inhibited the *in vitro* proliferation of tumour cells from MDV-infected chickens

(Lam and Linna, 1979), and, after activation with lipopolysaccharide or conditioned medium of activated spleen cells, macrophages kill MDCC cells (Qureshi and Miller, 1991).

Macrophages from MDV-infected chickens can have a marked suppressive effect on the *in vitro* proliferation of splenocytes (Lee *et al.*, 1978). The phenomenon appears to coincide with the transient immunosuppression occurring at the end of the cytolytic phase of infection. Susceptible chickens that developed MD lymphomas manifested a marked depression in lymphocyte proliferation responses consistent with the permanent immunosuppression at this time. However, lymphocytes from chickens that survived infection with MDV or from HVT-vaccinated chickens, with or without MDV challenge, exhibited enhanced proliferative T-cell responses to mitogens (Lee *et al.*, 1978). Schat and Markowski-Grimsrud (2001) explained these macrophage-induced suppressive effects as being due to the release of NO, and suggested that this inhibition could have a beneficial effect on the host by preventing the uncontrolled proliferation of T cells.

Role of natural killer cells

Natural killer (NK) cells are an important population of innate immune cells that are non-adherent and non-phagocytic lymphocytes and are fundamental in defences against cytopathic viruses, particularly herpesviruses, and tumour cells (Cerwenka and Lanier, 2001). They represent a major population of lymphocytes that do not express clonally distributed receptors for antigens like classical B and T cells but share a number of common features with T lymphocytes, particularly the ability to release and respond to cytokines and chemokines. The function of NK cells is regulated by a balance between signals transmitted by activating receptors, which recognize ligands on tumours and virus-infected cells, and inhibitory receptors specific for major histocompatibility (MHC) class I molecules on their targets (Cerwenka and Lanier, 2001). Under normal conditions NK cells are mostly restricted to the peripheral blood, spleen and bone marrow, but they rapidly migrate to sites of inflammation in response to various chemo-attractants. There is growing evidence that NK cells play key roles in MD, contributing to genetic resistance, vaccinal protection and anti-tumour responses.

Sharma and Coulson (1979) described a population of cells isolated from the spleens of uninfected chickens that behaved like NK cells and could lyse target cells from the MDCC MSB-1. The cytotoxic activity of these NK-like cells was found in spleens of uninfected chickens of several different inbred lines and increased with age, being greatest when chicks were 7 weeks old. Moreover, transfer of spleen cells from uninfected 8-week-old donor chicks protected day-old recipient chicks that were challenged with MD-transplantable tumour cells (Lam and Linna, 1979) suggesting this NK-like population could be involved. The cell line LSCC-RP9, derived from a tumour induced by the Rous-associated virus-2, was found to be a much better target cell for measuring chicken NK-like activity than MDCCs (Sharma and Okazaki, 1981), and the cells display-ing this activity were found to be neither B nor T lymphocytes. Sharma and Okazaki (1981) also demonstrated that NK-like activity increased about 1 week after infection with MDV in both resistant and susceptible chicken genotypes. NK-like cytotoxicity was greater in the resistant chickens. Moreover, vaccin-ation of chicks with HVT and/or a serotype 2 MDV increased NK cell activity with the maximal effect occurring at 7 days after vaccination as well as protecting against challenge with virulent MDV (Heller and Schat, 1987). Quere and Dambrine (1988) also reported that after infection with virulent MDV, splenic NK-like cytotoxicity was increased and neonatal vaccination caused an earlier increase in this NK-like activity.

Collectively, these data suggest NK cells have an important role in protective immunity against MDV and are likely to be most active in the cytolytic/replica-tive phase of infection. Cells with NK-like activity have also been isolated from MD tumours (Sharma, 1983), and their cytotoxicity against the LSCC-RP9 tar-get was greater than that of cells isolated from spleens of the same birds. This NK-like cytotoxicity was greater when these cells were isolated from regres-sive tumours than from progressive tumours.

Cells with NK-like cytotoxicity have been isolated from chicken spleen cell preparations cultured with conditioned medium from mitogen-activated spleno-cytes (Loeffler et al., 1986). These cells were capable of lysing LSCC-RP9 tar-gets as well as a number of MDCCs. Moreover, cytotoxicity directed against LSCC-RP9 was inhibited by the addition of MDCC cells to the cytotoxic assay, suggesting the killing mechanisms are common to both targets (Heller and Schat, 1987). Like their mammalian counterparts, the chicken NK-like cells are larger than B and T lymphocytes, contain azurophilic granules and have a large granular appearance (Schat et al., 1986). Interestingly, Keller et al. (1992) reported that the MDCC, JMV-1, secretes soluble factors that enhance in vitro NK-like cytotoxicity against LSCC-RP9. This suggests that some MD tumour cells may be capable of releasing cytokines that modulate NK cell activity. The nature of these cytokines has not yet been identified, but this could readily be done using modern reagents.

NK cells could play an important role in MDV infection, as they do in other herpesvirus infections (Cerwenka and Lanier, 2001). Garcia-Camacho et al. (2003) reported NK-like activity in MD-resistant line-N chickens was increased in the first 2 weeks after MDV infection and more sustained than in the geneti-cally susceptible line-P chickens. They concluded that NK cells in the resistant line play a key role in destroying MDV-infected cells to more effectively con-trol its spread to other cells. Genetic resistance to MD in line N is associated with the B^{21} haplotype of the chicken MHC, whereas line P possesses the B^{19} haplotype (see Chapter 9). However, genes outside of the chicken MHC also make an important contribution to MD resistance. Genetic mapping showed that a region on chicken chromosome 1 (MDV1 locus) is strongly associated with

MD resistance, and this region has synteny with the NK-cell receptor locus on mouse chromosome 6 and human chromosome 12 (Bumstead, 1998). In the mouse this genetic locus is strongly associated with resistance to cytomegalovirus infection and appears to function through effects on NK cells (see Chapter 9). It therefore seems highly likely that NK cells have an important role in controlling MDV infection; however, this role will not be elucidated until NK cells are more fully characterized and chicken NK cell markers become available.

Acquired immune responses to MDV

Role of antibodies

It has been estimated that the MDV genome could encode up to 100 viral proteins (see Chapter 9), although immunoprecipitation of MDV-infected cell lysates using convalescent antisera from MDV-infected birds identified only 35 viral proteins (van Zaane *et al.*, 1992). Antibodies to several MDV glycoproteins have been identified in convalescent antisera, including those recognizing gB (also known as Marek's B antigen), gE and gI, though not gD (Churchill *et al.*, 1969; Brunovskis *et al.*, 1992). The exact role of these antibodies is not certain, for MDV is a highly cell-associated herpesvirus and therefore strictly intracellular and thus it would be expected that antibody-mediated responses are less important in protection from MD than cell-mediated immune responses. This is not the case, however. Both passively and actively acquired antibodies have been implicated in protective immunity against MDV.

The chick naturally acquires maternal antibodies from the vaccinated or MDV-infected hen via the egg or, in the experimental situation, by adoptive transfer of immunoglobulins obtained from MD-vaccinated or convalescent chickens. In either case, antibodies delay the development of clinical signs of MD, providing some protection against MD morbidity, tumour formation and mortality (Chubb and Churchill, 1969; Calnek, 1972; Burgoyne and Witter, 1973; Lee and Witter, 1991). However, the presence of maternal antibodies has a deleterious effect on vaccine efficacy, particularly cell-free vaccines, administered either *in ovo* or to the neonate (Calnek, 1972; Sharma and Graham, 1982). The greater effect on cell-free vaccines is most likely due to neutralization of the vaccine virus. For instance, gB is expressed on the plasma membrane of MDV-infected cells, so it is likely that interference with vaccine protection occurs through the opsonization of cells bearing MDV-encoded surface glycoproteins.

HVT-vaccinated chickens make antibody responses to MDV, especially after challenge with field isolates of MDV, and levels of anti-MDV antibodies are maintained for life, perhaps because the chickens remain persistently infected with cell-associated HVT or MDV for their lifespan. Lee and Witter (1991) reported that antibody titres of hens, vaccinated with HVT as chicks, were not improved by hyperimmunization with inactivated oil-emulsion MDV vaccines.

They also observed that virus-neutralizing antibodies are not essential for protection against MD. Schat and Markowski-Grimsrud (2001) suggested that non-neutralizing antibodies could be protective by coating the surface of virus-infected cells and blocking viral antigen receptors on the surface of adjacent cells, so preventing virus spread in the cytolytic/replicative phase of infection.

Antibodies from either HVT- or MDV-vaccinated chickens also contribute to the cell-mediated destruction of MDV-infected cells through an antibody-dependent cell cytotoxic mechanism (Kodama *et al.*, 1979; Ross, 1980). Neither the effector cells nor the target antigens have been identified.

The role of the humoral immune response in the tumour phase of MD is even less certain. Dandapat *et al.* (1994) adopted the novel approach of making anti-idiotypic antibodies against antibodies that recognized Marek's disease tumour surface antigens (MATSA). Anti-idiotypic antibodies recognize the antigen-binding domain (idiotype) of the antibody in question, and so can mimic the original antigenic determinant present on the tumour cell. Chicks vaccinated with anti-idiotypic antibodies expressed lower levels of MATSA$^+$ cells and were partially protected from clinical MD after challenge with MDV. Recently we have shown that MD tumour cells over-express CD30, a homologue of the Hodgkin's lymphoma antigen (Burgess and Davison, 2002; Chapter 8). A specific anti-CD30 immunoglobulin response was identified after genetically resistant chickens were hyperimmunized with oncogenic MDV, suggesting anti-tumour humoral immunity is implicated in the regression of CD30^{hi} lymphomas. The inferences from these two studies are that (1) anti-tumour immune responses are induced after infection with MDV; and (2) antibodies against tumour-specific antigens could contribute to the anti-tumour responses (see Chapter 8).

Role of cell-mediated immune responses

Anti-viral immunity against herpesviruses is principally mediated by CD8$\alpha\beta^+$cytotoxic T lymphocytes (CTL) and CD4$^+$ helper cells that secrete cytokines. The same is likely with MDV; however, because of the lymphotropic and highly cell-associated nature of the virus which readily reactivates *in vitro*, it has been difficult to demonstrate clearly the role of these cells using classical immunological techniques such as adoptive cell transfer, proliferation assays or depletion of specific lymphocyte subpopulations. Morimura *et al.* (1998) demonstrated that monoclonal depletion of either CD4$^+$ or CD8$^+$ lymphocytes prevented HVT-vaccinated chickens from developing MD, although in the case of the CD4$^+$ cell depletion this may have been due to removal of target cells for the virus to infect and transform. Depletion of CD8$^+$ cells led to higher MDV titres in CD4$^+$ cells, suggesting CD8$^+$ cells exert an important anti-viral effect that influences the course and outcome of the disease.

Measurement of cytotoxic activity of effector CD8$^+$ T cells from MDV-infected chickens using chromium release assays has proved to be remarkably difficult because of problems obtaining suitably infected target cells (see Schat and Markowski-Grimsrud, 2001). MDCCs make poor CTL targets in syngeneic killing systems, although they are readily lysed in allogeneic assays. To circumvent these difficulties, Schat and co-workers (reviewed in Schat and Markowski-Grimsrud, 2001) developed a heterologous system to investigate CTL responses to MDV based on reticuloendothelial virus-transformed cells stably transfected with specific MDV genes and expressing their peptides. Individual cell lines transfected with a number of different MDV genes, including ICP4 and ICP22 (immediate-early), pp38 (early), Meq (early oncoprotein) and gB (late), were developed from chicken haplotypes that were MD-resistant (B^{21}/B^{21}) and MD-susceptible (B^{19}/B^{19}). The amounts of killing of these targets were low compared to mammalian CTL assays. However, stronger CTL killing was directed against gB- and pp38-transfected cells, weaker killing with Meq-transfected cells, but no lysis with ICP22-transfected cells. There was a differential effect with the killing of ICP4-transfected cells for the susceptible B^{19}/B^{19} line was unable to develop a CTL response, while the resistant B^{21}/B^{21} line developed a strong response. Omar and Schat (1996) speculated that MD resistance in the B^{21}/B^{21} haplotype is related to the specific recognition of the ICP4 immediate-early protein, which allows earlier detection and removal of cells engaged in productive MDV infection. Using *in vitro* depletion techniques the phenotype of the effector cells was confirmed as $CD8^{+}$ $TCR\alpha\beta1^{+}$ – in other words, classical CTL. Depletion of $CD4^{+}$, $TCR\alpha\beta2^{+}$ or $TCR\gamma\delta^{+}$ T cells had no effect.

A novel approach to investigating *in vitro* cell-mediated immune responses to MDV (Ross, 1977) used a plaque-reduction assay with target lymphocytes from MDV-infected chickens and effector lymphocytes from syngeneic chickens co-cultured over a reporter cell monolayer of chicken embryo fibroblasts. Plaque reduction only occurred when cells were obtained from immune chickens that had been inoculated with an attenuated strain of MDV. Cells from uninfected birds caused no reduction in plaque numbers. The effector cells were shown to be T cells but not B cells. The system used syngeneic effector and target cells, but subsequently it was shown that plaque reduction could also occur with an allogenic combination. We have confirmed and extended these observations, demonstrating that direct cell:cell contact is essential for plaque reduction to occur. Some of the effector cells have the $CD8\alpha\beta$ phenotype, but there is another, as yet unidentified, effector population that expresses neither $CD8\alpha$ nor the $TCR\gamma\delta$ receptor (B. Baaten, C. Butter and T. F. Davison, unpublished observations). It is tempting to speculate that these effector cells, which are clearly not CTL, are NK-like cells. The phenotype of these cells is currently being investigated.

Whilst it is generally accepted that CTL responses play a major role in reducing MDV replication, transmission and persistence, the role of CTL in anti-tumour immunity is more controversial. Payne *et al.* (1976) proposed a 'two-step' hypothesis consisting of anti-viral and later anti-tumour immune responses. The two-step hypothesis explains why MD proliferative lesions regress when older chickens are infected with MDV, and why lesion regression in MD-resistant chickens occurs at the same time as lymphomas are developing in MD-susceptible chickens (Burgess *et al.*, 2001). Schat (1991) questioned the two-step hypothesis, arguing that MDV can reactivate in tumour cells and MDV antigens may be expressed on a small number of cells, to stimulate anti-viral rather than anti-tumour responses. This could explain why chickens immunized with glutaraldehyde-fixed MDCC cells became partially protected after challenge with MDV, though better protected when immunized with MDV-infected kidney cells that express much higher levels of virus (Powell, 1975). It is impossible to differentiate anti-tumour from anti-viral immune responses because of the likelihood of at least a small number of tumour cells, or cell lines derived from them, expressing some MDV antigens. Recently we have shown that MD tumour cells over-express the CD30 antigen on their cell surface. We immunized chickens against CD30 using DNA and/or recombinant vaccines that expressed chicken CD30 and in one out of three experiments were able to show a delay in clinical MD and reduced tumours (see Chapter 8). These data indicate that anti-tumour immune responses may occur in MD just as reported for other oncogenic herpesviruses.

Cytokine responses during the cytolytic phase

Until recently the role of cytokines in the pathogenicity of, and immune responses to, MD had been poorly understood. Studies in the 1970s showed that interferons (IFNs) inhibit *in vitro* replication of MDV. Xing and Schat (2000b) investigated, both *in vitro* and *in vivo*, the effects of MDV infection on transcription of a number of cytokines in splenocytes from resistant (B^{21}/B^{21} haplotype) chickens. IFN-γ transcription was increased from as early as 3 dpi until at least 15 dpi, the time the experiment terminated. There was also up-regulation of IL-1β and iNOS after 6 dpi. They proposed that IFN-γ plays a pivotal role in the early pathogenesis and immune responses to MDV infection.

The development of a comprehensive panel of chicken cytokine reagents (reviewed in Secombes and Kaiser, 2003) made it possible to investigate mechanisms controlling the innate and acquired immune responses in MD. Chicken orthologues of the Th1 cytokines IFN-γ, interleukin (IL)-2, IL-12 p40 and IL-18, the pro-inflammatory cytokines IL-1β and IL-6, and IL-15 which is closely related to IL-2, as well as the chemokine IL-8, have been cloned and sequenced. Information on genomic sequences and gene structures of these cytokines has been used to design probes and primers for real-time quantitative RT-PCR assays to follow changes in cytokine transcription during the course of MD in resistant and susceptible chicken genotypes.

Kaiser *et al.* (2003) reported production of cytokine during the course of MDV infection in four inbred lines of chicken; two resistant (lines 6 (B^2/B^2) and N (B^{21}/B^{21})) and two susceptible (lines 7 (B^2/B^2) and P (B^{19}/B^{19})). IFN-γ mRNA was expressed by splenocytes from all infected birds between 3 and

10 dpi, associated with increasing MDV loads, in broad agreement with the results of Xing and Schat (2000b). For other cytokines, differences between lines were only seen for IL-6 and IL-18, with splenocytes from susceptible birds expressing high levels of both transcripts during the cytolytic phase of infection, whereas splenocytes from resistant birds expressed neither transcript. These results suggest that these two cytokines could play a crucial role in driving immune responses. In susceptible lines this results in lymphomas, but in resistant lines latency is maintained.

From a more limited time-course study following MDV infection or vaccination, Djeraba *et al.* (2002a) reported increased levels of IFN-γ again broadly agreeing with Xing and Schat (2000b). These studies contrast with earlier reports that after MDV infection resistant genotypes have higher levels of IFN than susceptible genotypes (Hong and Sevoian, 1971). Furthermore, there are conflicting reports on IFN production following HVT vaccination, suggesting differences could be caused by virus strain (Hong and Sevoian, 1971; Djeraba *et al.*, 2002a). Djeraba *et al.* (2002a) also reported increased levels of two CC chemokines, MIP1β and K203, in splenocytes from B^{21}/B^{21} and B^{13}/B^{13} haplotypes following MDV infections. Taken all together, these results indicate that MDV infection induces cytokines characteristic of a T_H1 immune response that might be expected following a viral infection.

The role of cytokines in the induction and maintenance of latency

The role that cytokines play in inducing and maintaining latent infection is one that merits reinvestigation with the modern reagents now available. Buscaglia and Calnek (1988) identified two soluble factors in the conditioned media from concanavalin A-activated splenocytes that could maintain MDV latency in splenocyte cultures from MDV-infected chickens. One of these was identified as type I IFN, the other, shown not to be IL-2, was named 'latency maintaining factor' (LMF). Interestingly, when conditioned medium was obtained from splenocytes activated by a mixed lymphocyte response LMF was not detected (Volpini *et al.*, 1996). Type I IFN was largely responsible for maintaining MDV in a latent state in MDV-transformed cell line cells. The precise nature of LMF and the phenotype of the cells producing it have yet to be identified.

Cytokine production by tumour cells

There are few reports on the production of cytokines by MDCCs or tumour cells. Bumstead and Payne (1987) reported that MDCC cells secrete a factor that suppresses the *in vitro* proliferation of mitogen-activated splenocytes. This factor was trypsin-sensitive and heat-resistant, with a MW of 20 kDa, but was not further identified. We investigated the expression of IL-2, IL-15 and IFN-γ transcripts by cell lines using quantitative RT-PCR assays, comparing lines

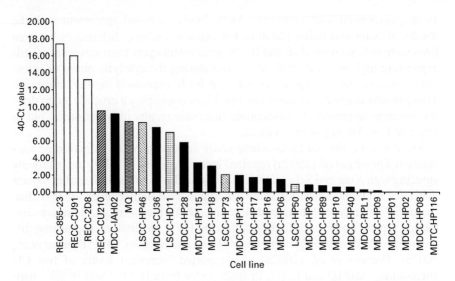

Figure 10.1 Measurement of IFN-γ mRNA transcripts isolated from 10^6 cells from a range of different virus-transformed cell lines using quantitative RT-PCR (Taqman). The chicken cells had been transformed with lymphoid leukosis (LSCC – stippled bars), MDV (MDCC – black bars) or reticuloendotheliosis virus (RECC – white bars). Turkey cells had been transformed with MDV (MDTC). RECC-CU210 expressed MDV proteins and is derived from the RECC-CU91 line (striped bar). Two macrophage cell lines, LSCC-HD11, and the MQ line (wavy lines), are included. The data are expressed as Ct values, each unit approximating a log 2 difference.

transformed with different oncogenic viruses such as MDV, lymphoid sarcoma viruses (LSV) and reticuloendotheliosis virus (REV). Flow cytometric analysis confirmed that all the MDV-transformed cell lines were T lymphocytes, although many had lost a number of their cell-surface antigens. IL-2 transcript was not detected in any of the cell lines tested, and IL-15 was only detected in macrophage cell lines and RECC-2D8. Moderate or low levels of IFN-γ transcript were only detected in 4 of 16 MDCCs and 1 of 2 turkey cell lines (Figure 10.1). Interestingly, IFN-γ transcription was most prolific in the REV-transformed cells and in one line that had been transfected with MDV transcripts (RECC-CU210, a kind gift from Dr M. Parcells) (see Schat and Markowski-Grimsrud, 2001); expression of the IFN-γ transcript was markedly lower than in the parent cell line (RECC-CU91). In addition, IL-12 p40 has been shown to be expressed in 10 out of 12 MDCCs (Balu and Kaiser, 2003).

Role of the MDV-encoded CXC chemokine

IL-8 is an important chemoattractant for T cells, amongst others. The MDV003 and MDV078 genes encode a CXC chemokine described in the literature as a homologue of IL-8 (MDV vIL-8: Liu *et al.*, 1999; see Chapter 4), despite IL-8-like activity not having been demonstrated. Xing and Schat (2000b) reported that IL-8 was expressed in splenocytes from MDV-infected birds after 3 dpi. This led them to speculate that the production of this CXC chemokine during the lytic infection of B cells might attract T cells to the areas of virus replication and, further, that the observed early expression of IFN-γ following MDV infection should stimulate the expression of IL-8 receptors on T cells. There are several problems with this proposition. First, there was no evidence of MDV-induced changes in chicken IL-8 expression in splenocytes from either MD-resistant or MD-susceptible chicken lines (Kaiser *et al.*, 2003). More importantly, although the MDV CXC chemokine (MDV vIL-8) has high amino acid identity with human IL-8 and chicken IL-8/CAF, it has higher identity with a recently identified B-lymphocyte chemoattractant. There are several important differences between the vIL-8 and known IL-8s. In mammals, CXC chemokines can be subdivided into two groups. Many possess an ELR motif immediately preceding the first cysteine residue, including IL-8. Others, such as PF-4 and the B-lymphocyte chemoattractant, lack this ELR (Legler *et al.*, 1998). In general, ELR$^+$ CXC chemokines are involved in angiogenesis, whereas ELR$^-$ CXC chemokines are not. The MDV vIL-8 lacks an ELR motif. In man, the genomic structure of most CXC chemokines consists of four exons and three introns (including IL-8), whereas the genes for PF-4 and NAP-2 comprise only three exons and two introns. The MDV vIL-8 gene also has three exons and two introns. It is therefore better that MDV vIL-8 be called vCXC rather than as a vIL-8. It is tempting to hypothesize that the vCXC is in fact a B-lymphocyte chemoattractant, which functions early in infection to attract B cells that can then be cytolytically infected by MDV. The precise role of the vCXC in subverting the immune response remains to be elucidated.

Immunity to MDV induced by vaccination

Vaccination against MDV was the first example of the successful mass administration of a vaccine that protects against a naturally occurring virus-induced cancer. Live vaccines are administered either to day-old-chicks or to the 18-day-old embryo (see Chapter 13). However, although MD vaccines target MDV replication in the cytolytic phase (Figure 10.2) and prevent lymphoma development, they do not prevent infection and replication of pathogenic strains of MDV. Although the innate and acquired immune responses have not been fully elucidated, it is highly likely that they are similar to those invoked by pathogenic strains. Attenuated strains of serotype 1 MDV are most antigenically identical to pathogenic strains, and probably induce more effective immune responses (see Chapter 3). Recent evidence suggests that a number of MDV proteins are protective, especially the glycoproteins such as gB that protect against MD

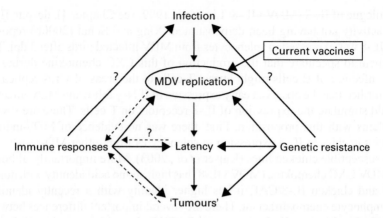

Figure 10.2 Schematic diagram showing major stages in the course of MD and those that are targets for immune responses, vaccines and genetic resistance. The immuno-suppressive effects of MDV are indicated by dotted lines. Question marks indicate those pathways where interactions are possible but require confirmation.

when administered as a recombinant vaccine (FPV-gB1) in a fowlpox vector (see Chapter 13).

Moreover, HVT has been used as a useful vector for delivering antigens from other pathogens such as Newcastle disease virus (NDV) and protection against both MD and Newcastle disease (see Chapter 13). HVT has the advantage that it persists for long periods in the chicken, as shown by the persistence of anti-bodies. Clearly there remains much to be done to identify the important MDV antigens that stimulate protective humoral and cell-mediated immune responses.

Cytokines could have an important role as immunomodulators and vaccine adjuvants, although to date only one cytokine, avian myelomonocytic growth fac-tor (MGF), has been tested for its efficacy in controlling infection with MDV *in vivo* (Djeraba *et al.*, 2002b). Delivered via a live fowlpox vector, chickens treated with MGF had prolonged survival time, lower viraemia and lower tumour inci-dence following challenge with the vvMDV strain, RB-1B. In the spleens of MGF-treated birds, there was increased induction of iNOS, IFN-γ and K203 expression. Systemic NO production was also increased, as were numbers of systemic monocytes. MGF may also have potential as a vaccine adjuvant with MDV, as it improved the partial protection given by vaccination with HVT, pre-venting mortality and reducing tumour development (Djeraba *et al.*, 2002b). The potential of other cytokines, particularly those involved in driving T_H1 responses, to act as vaccine adjuvants with MDV remains to be investigated.

Recombinant DNA technologies now offer the opportunity to develop and exploit molecularly-defined vaccines. However, to improve on conventional vac-cines it will be necessary to gain a fundamental understanding of the innate and adaptive immune responses to MDV as well as the mechanisms by which current vaccines protect. Conventional MD vaccines stimulate innate responses as well

Colour plates

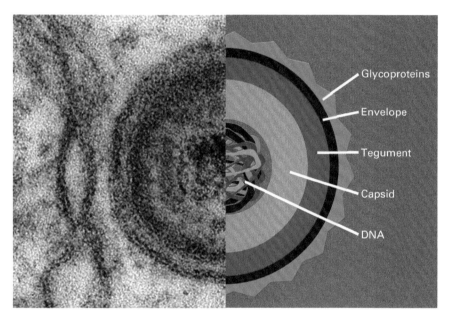

Plate 1 Electron micrograph showing an enveloped MDV in a chicken embryo fibroblast that had been infected with BAC20. This BAC clone was constructed from vv+MDV (584 strain) that had been attenuated by 80 passages *in vitro* (Schumacher *et al.*, 2000). Alongside is a schematic diagram showing the structure of the enveloped virus. (Electron micrograph reproduced courtesy of Daniel Schumacher, Cornell University Veterinary College.)

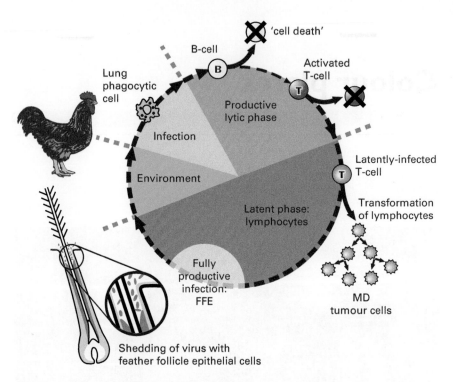

Plate 2 Schematic drawing depicting different stages in the cycle of MD pathogenesis. Not all of the stages are shown, but full details can be found in the text.

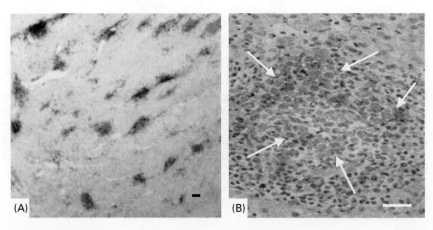

Plate 3 (*Cont'd*).

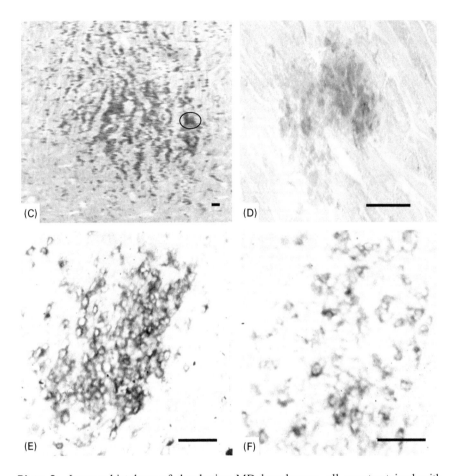

(C) (D)

(E) (F)

Plate 3 Immunohistology of developing MD lymphomas, all counterstained with haematoxylin and eosin. Bars = 100 μm. (A) Developing MD lymphomas from line MD-infected line 7_2 chicken heart, 21 days post-infection (dpi). MD-lymphomas form as multiple foci, each focus is composed of mainly CD4$^+$ T cells (red) associated with fewer myeloid (blue, mAb = CVI-ChNL-68.1) cells, which mainly surround, although a few are present within, each developing MD lymphoma focus. (B) Developing lymphoma in line 7_2 chicken liver, 21 dpi; CD30^{hi} cell aggregations (mAb = AV37) are present within MD lymphomas (arrowed). (C) MD lesion in sciatic nerve of 21-day-old 'MD-resistant' line 6_1 chicken at 7 dpi. Brown-stained cells are CD4$^+$; (D) cluster of CD30^{hi} cells from serial section of lesion in C (circled). (E & F) Differential expression of CD30 with different numbers of CD8$^+$ cells in lesions. Both lesions were on the same heart section of a line 7_2 chicken. Red = CD30$^+$ cells, blue = CD8$^+$ cells. (E) There are few CD8$^+$ cells in this lesion, CD30 expression is 'high' (intense red staining). (F) There are relatively many CD8$^+$ cells in this lesion and CD30 expression is 'low' (pale red staining). (From Burgess *et al.*, 2001, with kind permission.)

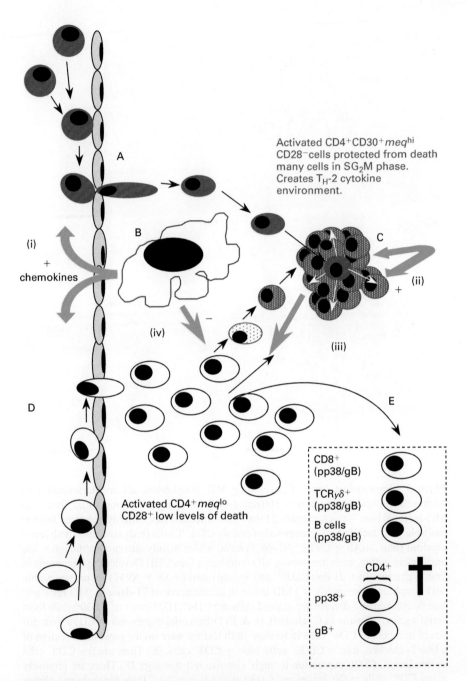

Activated CD4+CD30+ *meq*^hi CD28− cells protected from death many cells in SG₂M phase. Creates T_H-2 cytokine environment.

(i)
+
chemokines

A

B

C

(ii)
+

(iv)

−

(iii)

D

E

Activated CD4+ *meq*^lo CD28+ low levels of death

CD8+
(pp38/gB)

TCR_γδ+
(pp38/gB)

B cells
(pp38/gB)

CD4+

pp38+

gB+

†

Plate 4 Cellular model for Marek's disease lymphomagenesis. CD4+ cells, neoplastically-transformed and expressing high levels of the CD30 migrate into tissue stroma across endothelia (A, solid red cells). The stimulus for this migration may be

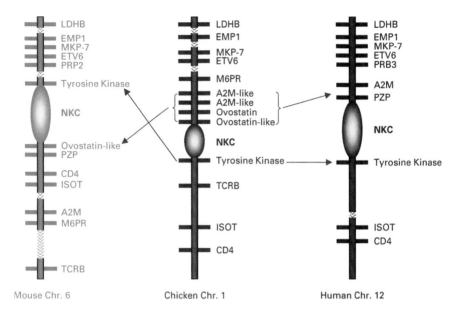

| Mouse Chr. 6 | Chicken Chr. 1 | Human Chr. 12 |

Plate 5 Comparison of the gene order spanning the NKC region on mouse chromosome 6, chicken chromosome 1 and human chromosome 12, kindly provided by Mark Clendenning, IAH. The genes shown (generally in the order top to bottom) are: LDHB, lactate dehydrogenase B; EMP1, epithelial membrane protein-1; MKP-7, MAP kinase-7; ETV6, ETS variant gene 6; PRP2, proline-rich protein-2; PZP, pregnancy zone protein-2; M6PR, mannose-6-phosphate receptor; A2M, alpha-2 macroglobulin; TCRB, T-cell receptor β; ISOT, isopeptidase T; and CD4, the gene encoding the CD4 T-cell antigen.

Plate 4 (*Cont'd*)   non-specific because of their activated state; and/or in response to chemokines (i) released from cells of the innate immune system such as macrophages (B); and/or in response to up-regulation of MHC class II/adhesion molecules or cytokines such as IFN by endothelial cells. The neoplastically-transformed CD30^hi cells clonally expand *in situ* (C) and release probably T_H2 cytokines, which have both autocrine (ii) and paracrine effects (iii). Latently infected , non-neoplastically-transformed (CD30⁻) lymphocytes (black) are attracted to the sites of the developing lesions by similar mechanisms to the neoplastically-transformed CD30 CD4⁺ cells (D). A small proportion of these cells become productively infected (boxed) and provide antigenic stimulus for a continued 'immune response' and lymphoid infiltration (E). CD30^hi neoplastically-transformed cells somehow escape this immune response. Infiltrating CD4⁺ cells may also potentially neoplastically-transform *in situ* and then over-express CD30. Macrophages (B), present in the lesions, as well as the resident neoplastically-transformed cells are also likely to influence the ongoing immune response in a paracrine manner (iv).

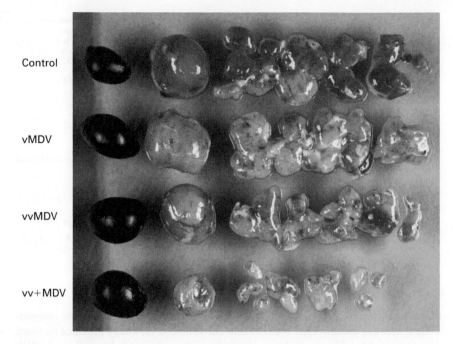

Plate 6 Lesions induced in lymphoid organs 8 days post-infection of 24-day-old P2a chickens with strains representing different pathotypes. (Reproduced courtesy of B.W. Calnek.)

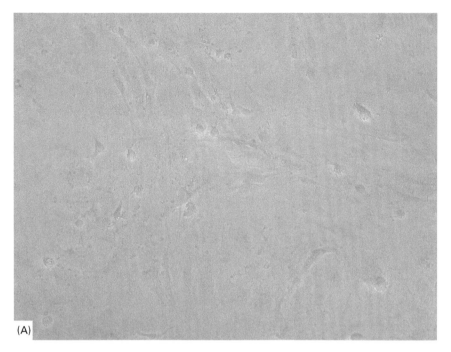

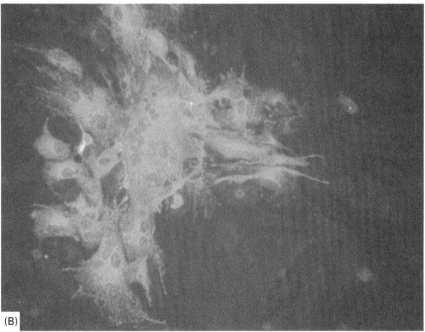

Plate 7 Cytopathic effect of very virulent MDV, EU1 strain, on DEF cells. (A) Phase contrast image of infected cells; (B) fluorescent image after staining with MDV-specific anti-pp38 chicken immunoglobulins and Alexa-Fluor-labelled goat anti-chicken immunoglobulins.

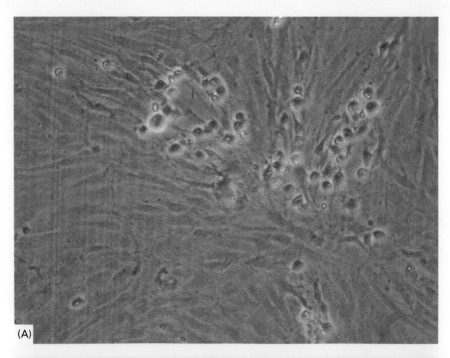

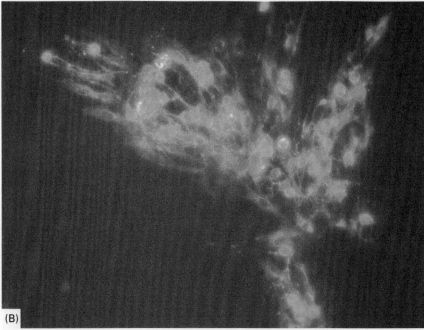

Plate 8 Cytopathic effect of vaccine MDV, CVI 988/Rispens strain, on CEF cells. (A) Phase contrast image of infected cells; (B) fluorescent image after staining with MDV specific anti-pp38 chicken immunoglobulins and Alexa-Fluor-labelled goat anti-chicken immunoglobulins.

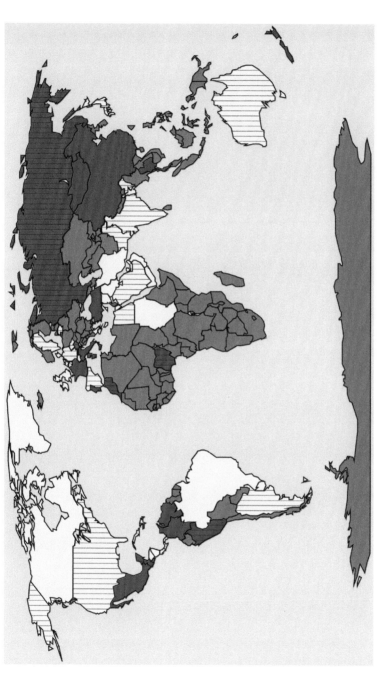

Plate 9 Map of the evolution of MD incidence. Colours represent the current existence of MD outbreaks as follows: red = current MD outbreaks; yellow = no current MD outbreaks; grey = data not available. The existence of MD outbreaks during 1990s is indicated by vertical stripes.

as stimulating both the B- and T-dependent immune systems eliciting antibody and cell-mediated immune responses. Not all of the responses are protective, and use of novel vaccines to polarize immune responses and improve protection could have a crucial effect in the fight against MDV evolving to greater virulence (see Chapter 14). Selection of pathogen molecules that stimulate protective responses, incorporation of genes that encode the host's own cytokine molecules to direct and enhance protective immune responses, or targeting immune responses to MD tumour antigens may be highly desirable for the future.

Conclusions

Immunity to MD involves a complex interplay of innate and acquired responses, the latter involving both cell-mediated (T_H1) and antibody (T_H2) responses. However, there is still much to be learned concerning the important target antigens of MDV and the protective host immune responses. MDV has a complex life cycle including a transient cytolytic (semi-productive) phase, latent and tumour phases, with life-long fully productive infection in the feather-follicle epithelium; so it seems likely that different immune responses will be important at different locations and in different stages of the infection. One of the important problems with MDV is that conventional vaccines have, so far, not been able to engender a sterilizing immunity. This is a fundamental drawback and one of the major factors contributing to the vaccination-induced pressure on MDV to evolve to greater virulence. Invoking protective immune responses that prevent MDV infection and/or replication and being shed from vaccinated chickens is perhaps the greatest challenge for immunology in the future (see Chapter 14).

References

Balu, S. and Kaiser, P. (2003). *J. Interferon Cyt. Res.*, **23**, 699–707.

Barrow, A. D., Burgess, S. C., Baigent, S. *et al.* (2003). *J. Gen. Virol.*, **84**, 2635–2645.

Brunovskis, P., Chen, X. and Velicer, L. F. (1992). In *4th International Symposium on Marek's Disease and XIX World's Poultry Congress, Amsterdam*, pp. 118–122. Ponsen and Looijen, Wageningen.

Bumstead, J. and Payne, L. N. (1987). *Vet. Immunol. Immunopathol.*, **16**, 47–66.

Bumstead, N. (1998). *Avian Pathol.*, **27**, S78–S81.

Burgess, S. C. and Davison, T. F. (2002). *J. Virol.*, **76**, 7276–7292.

Burgess, S. C., Basaran, B. H. and Davison, T. F. (2001). *Vet. Pathol.*, **38**, 129–142.

Burgoyne, G. H. and Witter, R. L. (1973). *Avian Dis.*, **17**, 824–837.

Buscaglia, C. and Calnek, B. W. (1988). *J. Gen. Virol.*, **69**, 2809–2818.

Buscaglia, C., Calnek, B. W. and Schat, K. A. (1988). *J. Gen. Virol.*, **69**, 1067–1077.

Calnek, B. W. (1972). *Infect. Immunity*, **6**, 193–198.

Calnek, B. W., Schat, K. A., Ross, L. J. N. *et al.* (1984). *Intl J. Cancer*, **33**, 389–398.

Cerwenka, A. and Lanier, L. L. (2001). *Nature Immunol. Rev.*, **1**, 41–49.

Chubb, R. C. and Churchill, A. E. (1969). *Vet. Record*, **85**, 303–305.

Churchill, A., Chubb, R. C. and Baxendale, W. (1969). *J. Gen. Virol.*, **4**, 557–564.

Dandapat, S., Pradhan, H. K. and Mohanty, G. C. (1994). *Vet. Immunol. Immunopathol.*, **40**, 353–366.

Djeraba, A., Musset, E., Bernardet, N. *et al.* (2002a). *Vet. Immunol. Immunopathol.*, **85**, 63–75.

Djeraba, A., Musset, E., Lowenthal, J. W. *et al.* (2002b). *J. Virol.*, **76**, 1062–1070.

Garcia-Camacho, L., Schat, K. A., Brooks, R. and Bounous, D. I. (2003). *Vet. Immunol. Immunopathol.*, **95**, 145–153.

Gupta, M. K., Chauhan, H. V., Jha, G. J. and Singh, K. K. (1989). *Vet. Microbiol.*, **20**, 223–234.

Haffer, K., Sevoian, M. and Wilder, M. (1979). *Intl J. Cancer*, **23**, 648–656.

Heller, E. D. and Schat, K. A. (1987). *Avian Pathol.*, **16**, 51–60.

Higgins, D. A. and Calnek, B. W. (1976). *Infect. Immunity*, **13**, 1054–1060.

Hong, C. C. and Sevoian, M. (1971). *Appl. Microbiol.*, **22**, 818–820.

Kaiser, P., Underwood, G. and Davison, F. (2003). *J. Virol.*, **77**, 762–768.

Kaleta, H. D. and Neumann, U. (1977). *Avian Pathol.*, **6**, 33–39.

Keller, L. H., Lillehoj, H. S. and Solonsky, J. M. (1992). *Avian Pathol.*, **21**, 239–250.

Kodama, H., Mikami, T., Inoue, M. and Izawa, H. (1979). *J. Natl Cancer Inst.*, **63**, 1267–1271.

Lam, K. M. and Linna, T. J. (1979). *Intl J. Cancer*, **24**, 662–667.

Lee, L. F. and Witter, R. L. (1991). *Avian Dis.*, **35**, 452–459.

Lee, L. F., Sharma, J. M., Nazerian, K. and Witter, R. L. (1978). *Infect. Immunity*, **21**, 474–479.

Legler, D. F., Loetscher, M., Roos, R. S. *et al.* (1998). *J. Exp. Med.*, **187**, 655–660.

Liu, J.-L., Lin, S.-F., Xia, L. *et al.* (1999). *Acta Virol.*, **43**, 94–101.

Loeffler, D. A., Schat, K. A. and Norcross, N. L. (1986). *J. Clin. Microbiol.*, **23**, 416–420.

Morimura, T., Ohashi, K., Sugimoto, C. and Onuma, M. (1998). *J. Vet. Med. Sci.*, **60**, 1–8.

Omar, A. R. and Schat, K. A. (1996). *Virology*, **222**, 87–99.

Payne, L. N., Frazier, J. A. and Powell, P. C. (1976). *Intl Rev. Exp. Pathol.*, **16**, 59–154.

Powell, P. C. (1975). *Nature*, **257**, 684–685.

Powell, P. C. and Davison, T. F. (1986). *Israel J. Vet. Med.*, **42**, 73–78.

Powell, P. C., Mustill, B. M. and Rennie, M. (1983). *J. Reticuloendothelial Soc.*, **34**, 289–297.

Quere, P. and Dambrine, G. (1988). *Ann. Rech. Vet.*, **19**, 193–201.

Qureshi, M. A. and Miller, L. (1991). *Poultry Sci.*, **70**, 530–538.

Ross, L. J. N. (1977). *Nature*, **268**, 644–646.

Ross, L. J. N. (1980). *CEC Publication EUR 6470*, Luxembourg.

Schat, K. A. (1991). *Poultry Sci.*, **70**, 1165–1175.

Schat, K. A. and Markowski-Grimsrud, C. J. (2001). *Curr. Topics Microbiol. Immunol.*, **255**, 91–120.

Schat, K. A., Calnek, B. W. and Weinstock, D. (1986). *Avian Pathol.*, **15**, 539–556.

Secombes, C. J. and Kaiser, P. (2003). In *The Cytokine Handbook* (eds A. Thomson and M. T. Lotze), pp. 57–84. Academic Press, London.

Sharma, J. M. (1983). *Vet. Immunol. Immunopathol.*, **5**, 125–140.

Sharma, J. M. and Coulson, B. D. (1979). *J. Natl Cancer Inst.*, **63**, 527–531.

Sharma, J. M. and Graham, C. K. (1982). *Avian Dis.*, **26**, 860–870.

Sharma, J. M. and Okazaki, W. (1981). *Infect. Immunity*, **31**, 1078–1085.

Toth, T. E. (2000). *Devel. Comp. Immunol.*, **24,** 121–139.

van Zaane, D., Brinkhof, J. M. A., Westenbrock, F. and Gielkens, A. L. J. (1992). *Virology*, **121,** 116–132.

Volpini, L. M., Calnek, B. W., Sneath, B. *et al*. (1996). *Avian Dis.*, **40,** 78–87.

Xing, Z. and Schat, K. A. (2000a). *J. Virol.*, **74,** 3605–3612.

Xing, Z. and Schat, K. A. (2000b). *Immunology*, **100,** 70–76.

Marek's disease immunosuppression

11

KAREL A. SCHAT

Department of Microbiology and Immunology, College of Veterinary Medicine, Cornell University, Ithaca, New York, USA

Introduction

Infection with Marek's disease virus (MDV) and subsequent development of Marek's disease (MD) is frequently associated with immunosuppression (reviewed in Schat and Markowski-Grimsrud, 2001; Witter and Schat, 2003; Parcells and Burgess, 2004), which is considered to be an integral aspect of the pathogenesis of MD (Calnek, 1986; Schat, 1987a). MDV-induced immunosuppression is often divided into an early transient phase during the initial cytolytic infection followed by a second permanent phase, when MDV replication is reactivated and tumours may develop. In this chapter the mechanisms and consequences of MDV-induced immunosuppression will be reviewed, as well as confounding factors such as co-infection with chicken infectious anaemia virus (CIAV) and other viruses infecting lymphoid tissues. This review is not exhaustive, and the reader is referred to other reviews for information from some of the older literature (Payne *et al.*, 1976; Schat, 1987b).

The term 'immunosuppression' is frequently used in avian health, but often without providing a definition. In this chapter, immunosuppression is defined as 'a state of temporary or permanent dysfunction of the immune response resulting from insults to the immune system and leading to increased susceptibility to disease', as originally proposed by Dohms and Saif (1984). These authors emphasized that immunosuppression includes damage to the immune system caused by pathogens and/or environmental factors such as social stress. In addition, they pointed out that increased susceptibility to disease is an integral aspect of immunosuppression. In the case of pathogen-induced immunosuppression it can become a vicious circle, increasing the disease associated with the immunosuppression-inducing pathogen. Immunosuppression can also cause increased susceptibility to other pathogens. If two or more immunosuppressive pathogens are involved it may become extremely difficult to differentiate between cause and effect, for example in MD vaccine breaks in commercial flocks.

Mechanisms of MD-induced immunosuppression

In order to examine the causes of MDV-induced immunosuppression after experimental infection it is essential to ensure that the MDV strain and the experimental chickens are free of other immunosuppressive pathogens. Over the last decade it has become clear that infection and subsequent replication of CIAV can influence the pathogenesis of MDV by preventing the development of MDV-specific cytotoxic T cells (CTL) (Markowski-Grimsrud and Schat, 2003; see later). Unfortunately, the importance of CIAV was not realized until recently and many of the papers describing MDV-induced immunosuppression do not mention the CIAV status of challenge virus and/or experimental chickens, which complicates the interpretation of some of the published information. In this review the causes of MDV-induced immunosuppression are divided into three categories:

1. Loss of lymphocytes as a consequence of virus replication
2. Virus-induced changes in the regulation of immune responses
3. Tumour cell-induced immunosuppression.

Immunosuppression caused by loss of lymphocytes

The early pathogenesis of MD has been reviewed by Calnek (2001) and in Chapter 6. Briefly, the initial replication of MDV starting between 2 and 4 days post-infection (dpi) occurs in B lymphocytes, which leads to activation of T lymphocytes, which then express major histocompatibility complex (MHC) class II antigens. These activated T lymphocytes, but not resting T cells, become infected. Approximately 7 to 10 dpi the infection may become latent and, depending on the virulence of the challenge strain, age at exposure and genetic resistance of the chickens, may remain latent or be reactivated with subsequent development of tumours. MDV-infected cells undergoing a productive-restrictive replication are destined to die, as is the case in all herpesvirus infections (Roizman, 1996). The term 'productive-restrictive' is used to indicate that MDV replication results in the production of cell-associated but not cell-free virus. In this review, the use of productive or lytic infection always refers to the productive-restrictive infection. Earlier studies had indicated that mostly CD4$^+$ cells expressing T cell receptor (TCR)$\alpha\beta$ can become lytically infected, but more recently it has been shown that TCR$\gamma\delta$ T cells and CD8$^+$ T cells can also undergo lytic infection (Burgess and Davison, 2002; Barrow *et al.*, 2003). Monocytes can also become infected, especially with very virulent (vv)+ strains, although it is not yet clear if this is an abortive or a productive infection (Barrow *et al.*, 2003).

MDV strains with increased virulence, especially the vv+MDV strains, may not establish latency when less virulent strains become latent under identical conditions (Calnek *et al.*, 1998; Jarosinski *et al.*, 2002). The consequences of prolonged cytolytic infection were clearly demonstrated by Calnek *et al.* (1998). Infection with vv+MDV strains (e.g. RK-1) caused significantly more damage to the lymphoid organs than infection with vMDV or vvMDV strains (Plate 6). Moreover, the lymphoid organs did not show signs of recovery in birds infected with vv+MDV at 14 dpi in contrast to those birds infected with v or vvMDV strains. One explanation for the increased cytolytic infection is a change in the virally-encoded IL-8 homologue gene and/or promoter/enhancer region, which could lead to increased attraction of activated T cells (Schat and Xing, 2000). Jarosinski *et al.* (2003) sequenced the vIL-8 gene and its promoter/enhancer region, and found no differences between the genes or promoter/enhancer region of MDV strains of low oncogenicity (e.g. CU2) and those of high oncogenicity (e.g. RK-1). Interestingly, major deletions were found in the promoter/enhancer regions of attenuated MDV-1 strains, which could explain the lack of *in vivo* infectivity for lymphocytes of attenuated viruses (Schat *et al.*, 1985). Several groups had suggested that pp38 may be involved in the cytolytic process. A pp38 deletion mutant was still able to induce MD lesions, albeit in lower numbers than the wild-type virus, but the deletion mutants did not induce a cytolytic infection (Reddy *et al.*, 2002). However, the organization and transcription of the pp38 region is rather complex (Parcells *et al.*, 2003; Li and Schat, unpublished observations), and additional studies are needed before pp38 can be implicated as the cause of cytolysis.

Apoptosis is the likely cause of MDV-induced cell death, as is the case in many virus infections (O'Brien, 1998). Herpesviruses often can trigger apoptosis during productive infections, but also have the ability to block apoptosis during latency or transformation (Derfuss and Meinl, 2002). Morimura *et al.* (1996) reported that CD4$^+$CD8$^+$ thymocytes became apoptotic during the lytic infection phase of MDV. However, it is not clear if all or some of the apoptotic thymocytes were infected with MDV or if the apoptosis was caused by MDV-induced alterations in the microenvironment of the thymus affecting thymocyte maturation as was suggested by Payne *et al.* (1976). This is an important unresolved question, especially in view of the finding that during the first cytolytic infection the majority of infected lymphocytes in the thymus are B cells (Schat *et al.*, 1983; Shek *et al.*, 1983; Calnek *et al.*, 1984). Apoptosis of CD4$^+$ T cells in peripheral blood has also been reported starting at 14 to 21 dpi, probably as part of the secondary cytolytic infection, although these apoptotic cells are not necessarily infected with MDV (Morimura *et al.*, 1995). As mentioned above, it has been shown that herpesviruses can both induce apoptosis and prevent its occurrence, which may be an important part of maintaining latency (Derfuss and Meinl, 2002). It is certainly feasible that MDV strains have been selected for a decreased anti-apoptotic or increased apoptotic ability during virus replication in vaccinated chickens, which could explain the prolonged cytolytic infection associated with the more virulent strains.

It is often assumed that the lymphocyte depletion, which results from increased cytolytic replication of MDV, will lead to enhanced oncogenicity. However, increased levels of cytolytic infection do not necessarily lead to an increase in tumour incidence. Jones *et al.* (1996) and Witter *et al.* (1997) reported that the MDV clone RM1 caused considerable damage to lymphoid organs, especially in the thymus, but that it was attenuated for oncogenicity. This lack of oncogenicity could be related to attenuation of RM1 in cell culture. Passage 39 of RM1 was compared to JM passage13, which is oncogenic, and JM passage 48, which is attenuated. The RM1 clone was derived from JM virus that had been co-cultured with reticuloendotheliosis virus (REV), and RM1 has a single copy of the REV long terminal repeat (LTR) inserted at the R_S/U_S boundary. The insertion of the LTR, a strong promoter, causes an increased transcription of a polycistronic message encoding the SORF2, US1, and US10 gene products. Based on unpublished data from Brunovskis and Kung, cited by Jones *et al.* (1996), it seems that the enhanced transcription increased the production of SORF2 protein, but not the US1 and US10 proteins. Evidence that SORF2, a unique MDV protein, may be important for lytic infection and subsequent immunosuppression came from studies by Parcells *et al.* (1995) in which mutant viruses with deletions in U_S sequence including SORF2 caused a significant reduction in the lytic infection without preventing tumour formation. The function of SORF2 is not clear, although it can interact with chicken growth hormone (GH) (Liu *et al.*, 2001). GH has been associated with the development of the normal thymic structure and recovery from thymic damage. Liu and co-workers suggested that excess SORF2 product binds GH, thus preventing repair of RM1-caused thymic damage.

In contrast to infection with MDV serotype 1, herpesvirus of turkeys (HVT) and serotype 2 (SB-1 strain), MDV causes few or no lesions in the lymphoid organs. Few splenocytes express viral antigens, although virus can be readily isolated from splenocytes by co-cultivation with chicken embryo fibroblasts (CEF) during the early infection period (Schat and Calnek, 1978; Calnek *et al.*, 1979; Fabricant *et al.*, 1982). However, Holland *et al.* (1994) demonstrated the presence of HVT glycoprotein B (gB) in peripheral blood mononuclear cells during that period. Absence of a strong cytolytic infection is of particular interest in light of the finding that HVT RSORF1 has two domains with significant similarity to the quail anti-apoptotic gene NR13 BH1 and BH2 domains (Kingham *et al.*, 2001). It will be important to determine the temporal expression of HVT RSORF1 during virus replication. Interestingly, MDV serotype 1 lacks a homologue of HVT RSORF1, which could in part explain the apoptosis during the early cytolytic infection. Ewert and Duhadaway (1999) also reported that HVT and SB-1, but not the attenuated R2/23 serotype 1 strain, inhibited apoptosis in the B cell line, DT40, although they did not identify potential anti-apoptotic genes in SB-1 or HVT.

Immunosuppression caused by changes in the regulation of immune responses

Lymphotropic viruses such as MDV do not only replicate in and cause the death of lymphocytes; virus replication may also cause more subtle changes in cytokine regulation or expression of antigens on the surface of lymphocytes. Many of these changes may be associated with the initiation of immune responses (see Chapter 10), but some changes may lead to immunosuppression and will be discussed below.

Nitric oxide, a dual role

Previous studies have shown that the production of nitric oxide (NO) by macrophages or perhaps other cells can be beneficial in the reduction of MDV replication *in vivo* (Chapter 10). Moreover, genetically resistant N2a chickens produce higher levels of NO than susceptible P2a chickens after infection with vMDV, JM strain (Jarosinski *et al.*, 2002). However, infection with vv+MDV strains produces significantly higher levels of NO *in vivo* than infection with vMDV strains, which could lead to NO-induced immunosuppression by the induction of apoptosis. NO can induce apoptosis by causing mitochondrial dysfunction in thymocytes (Bustamante *et al.*, 2000). It is possible that excessive production of NO during the cytolytic infection in the thymus is the cause of apoptosis in thymocytes. It is also interesting that vv+MDV causes more pronounced neurological symptoms than less virulent strains, especially in the resistant N2a line. Levels of inducible NO synthase (iNOS) mRNA in the brain are significantly higher after infection with vv+MDV than vMDV using qRT-PCR assays, which appears to be correlated with the level of cytolytic replication (K. W. Jarosinski, B. L. Njaa, P. H. O'Connell, K. A. Schat, unpublished observations). The induction of iNOS was independent of the production of interferon (IFN)-γ (Jarosinski *et al.*, 2002) suggesting that iNOS induction was caused by viral protein(s) as suggested for other virus infections (Akaike and Maeda, 2000). It is intriguing to note that Djeraba *et al.* (2002) also reported that resistant lines have an increased NO production compared to susceptible lines. This may be caused by an increased arginase production in susceptible chicken lines, which interferes with the iNOS pathway. Overall, these results suggest that NO may play an important anti-viral role in chickens infected with vMDV strains, but NO may have a detrimental impact after infection with vv+MDV strains, causing pathology especially in resistant chickens.

Cytokines

Infection of lymphocytes by MDV is expected to up-regulate as well as down-regulate the production of cytokines as part of the development of immune responses and as a consequence of the cytolytic infection. There are only a few publications showing up-regulation of specific cytokines based on semi-quantitative or qRT-PCR assays (Xing and Schat, 2000; Jarosinski *et al.*, 2002; Kaiser *et al.*, 2003). Thus far, there are no published data showing down-regulation of defined cytokines as a consequence of MDV infection. Perhaps the up-regulation of the pro-inflammatory cytokines IL-6 and IL-18 reported by Kaiser *et al.* (2003) in susceptible but not in resistant birds may aggravate lesions, leading to immunosuppression.

Down-regulation of lymphocyte surface molecules and other immuno-evasive mechanisms

Herpesviruses have developed a remarkable array of strategies to evade immune responses (Koszinowski and Hengel, 2002), but thus far few mechanisms have been identified for MDV. Studies using microarrays with RNA from MDV-infected CEF cultures showed that MHC class I molecules are weakly induced, while MHC class II and microglobulin β2 are moderately induced, suggesting that MDV infection results in an up-regulation of MHC expression in CEF (Morgan *et al.*, 2001). However, more detailed studies showed that increased expression of MHC class I was present in non-infected CEF, while infected cells had decreased levels of expression (Kent *et al.*, 2001). Hunt *et al.* (2001) and Levy *et al.* (2003) also showed that infection with MDV serotype 1 down-regulated MHC class I expression in the OU2 fibroblast cell line and CEF, respectively. Hunt *et al.* (2001) reported a high degree of down-regulation after HVT infection, while Levy *et al.* (2003) showed only a minimal effect after HVT infection. The decreased expression after infection with serotype 1 MDV is probably caused by changes in peptide transport or retention of class I MHC in the endoplasmic reticulum (Hunt *et al.*, 2001). Addition of a mixture of native chicken interferons containing IFN-α and IFN-γ (Heller *et al.*, 1997) restored the expression of class I antigen (Levy *et al.*, 2003). Induction of pp38 expression in MDV cell lines also resulted in a decreased expression of MHC class I, suggesting that pp38 could be involved in, or perhaps is responsible for, the down-regulation of MHC class I antigen (Hunt *et al.*, 2001).

Down-regulation of CD8 antigen on $CD4^+CD8^+$ and $CD4^-CD8^+$ lymphocytes in the thymus has been reported during the cytolytic infection (Morimura *et al.*, 1996) and at 21 dpi in splenic and peripheral blood lymphocytes. This was apparently caused by a decrease in transcripts for both the α and β chain genes (Morimura *et al.*, 1995). Decreased expression of CD8 may lead to decreased CTL activity, since CD8 interacts with the invariant part of the MHC class I molecule during antigen recognition. Decreased CTL function during the secondary cytolytic infection, when MDV-specific CTL are present (Schat and Markowski-Grimsrud, 2001), can have serious consequences for the control of MDV infections. It is currently not clear if the down-regulation is caused by infection of these cells or through indirect effects.

Immunosuppression caused by tumour cells

The development of tumours is often associated with (or preceded by) reactivation of virus replication, leading to additional destruction of lymphoid tissues and permanent immunosuppression. Early studies using mitogen stimulation assays suggested that tumour cells are immunosuppressive. Addition of JMV, a tumour transplant, cells or MDV-transformed lymphoblastoid cells to normal spleen cells inhibited the proliferative response to mitogens (Theis, 1981; Quere, 1992). It was found that supernatant fluids from MDV-infected spleen cells cultured *in vitro*, MD lymphoblastoid cell lines, and spleen cells infected *in vitro* with MDV were also able to suppress mitogen stimulation of normal spleen cells (Theis, 1977; Bumstead and Payne, 1987). Although the factor(s) responsible for the inhibition of mitogen response were not definitively identified, it was noted by Bumstead and Payne (1987) that these suppressive factor(s) could induce prostaglandin E_2 (PGE_2). The importance of this finding is not clear at the present time. Tumour cells can also be immunosuppressive by expressing inappropriate antigens on the surface. One of these antigens, chicken foetal antigen, interferes with natural killer (NK) cell activity (Ohashi *et al.*, 1987).

Several potentially immuno-evasive mechanisms have been described for MD tumour cells. MD tumour cells may have a decreased expression of MHC class I antigens after activation, resulting in viral antigen expression (Hunt *et al.*, 2001). Based on sequence data of some MDV strains and RT-PCR analysis in tumour cell lines, Shaw *et al.* (2001) suggested that MHC class I-like decoy molecules are produced, although this observation needs confirmation. Recently it has also been shown that CD28 is down-regulated on lymphoma cells *in situ* (Burgess and Davison, 2002), but not in tumour cell lines (Parcells and Burgess, 2004). The down-regulation of CD28 at the transcriptional level has potentially important consequences because it is an important co-stimulatory molecule for T-cell activation, and the absence of CD28 may be one of the causes of immuno-evasion of the tumour cells. Another potentially important antigen in the deregulation of immune responses to MD tumour cells is the high level of expression of CD30 on tumour cells (S. C. Burgess *et al.*, personal communication), which may cause a switch toward a T_H2 response. This is interesting because MD cell lines derived from the local lesion model (Calnek *et al.*, 1989a) can produce NO, often together with IFN-γ, a typical T_H1 response, while cell lines derived from tumours are negative for both (Z. Xing, K. W. Jarosinski, P. H. O'Connell, K. A. Schat, unpublished). Perhaps the balance between arginase/iNOS production, as suggested by Djeraba *et al.* (2002), is important in this context if excess CD30 production causes a shift to a T_{H2} response resulting in increased arginase production and a decreased NO production. This may have important consequences if tumour cells are susceptible to NO-induced killing or growth inhibition.

Confounding factors in MDV-induced immunosuppression

One of the major problems in assessing if MDV is indeed the cause of immuno-suppression, especially in commercial flocks, is the potential impact of stress (Gross, 1972) or the presence of other viruses infecting lymphoid tissues. This section will briefly examine the impact of CIAV, infectious bursal disease virus (IBDV), reticuloendotheliosis virus (REV) and reovirus on MDV infection. One problem, especially with older studies, is that it is not always clear if the IBDV, REV and reovirus isolates were free of CIAV.

Chicken infectious anaemia virus

Chicken infectious anaemia virus has long been associated with MDV vaccine breaks or aggravation of MD in experimental infections, and is probably the most important confounding pathogen (reviewed by Schat, 2003). In a recent study in Australia, 11 of 49 MDV isolates obtained from flocks experiencing MD breaks were contaminated with CIAV (G. A. Tannock, personal communication). In another study, in Germany, MD breaks in older birds were also linked to the presence of CIAV infections (Fehler and Winter, 2001). Experimental studies, however, do not always provide a clear picture. For example, co-infection of CIAV with RB1B, a vvMDV strain, or 584A, a vv+MDV strain, exacerbated MD mortality compared to infection with MDV alone (Miles *et al.*, 2001). On the other hand, Jeurissen and de Boer (1993) indicated that the effects may be related to the challenge dose of MDV, with a high dose of MDV reducing the impact of CIAV co-infection. This is perhaps due to the lack of target cells for CIAV as a consequence of the MDV lytic infection. Vaccination against MD reduced the impact of CIAV infection when challenged with vMDV plus CIAV 6 to 7 days later (Zanella and Dall'Ara, 2001). This is perhaps due to the protection against MDV-induced cytolytic infection of B cells allowing the development of antibody responses to CIAV, thus curtailing the replication of CIAV. The most likely explanation for CIAV-induced immunosuppression was recently provided by Markowski-Grimsrud and Schat (2003). They showed that CIAV infection can abolish CTL responses to a second pathogen when both pathogens are simultaneously replicating. This finding has major consequences for the interpretation of many of the older studies on the consequences of MDV-associated immuno-suppression. For example, Jakowski *et al.* (1970) described haematopoietic lesions after experimental MDV infection, but it was shown much later that their MDV strain was contaminated with CIAV (C. R. Wellenstein, personal communication 1989, quoted by Schat, 2003).

Infectious bursal disease virus

Infectious bursal disease virus can cause severe depression of antibody responses by depletion of immunoglobulin-producing cells. IBDV infection also can decrease T_H cell functions, probably through macrophage-produced antiproliferative cytokines and NO (Sharma *et al.*, 2000). Yet the actual effects of IBDV on MD are not clear. IBDV infection in IBDV-maternal antibody-negative chickens interfered with the efficacy of HVT vaccination (Jen and Cho, 1980; Sharma, 1984). However, dual infection with IBDV and MDV, or IBDV infection followed by MDV at 14 days, actually reduced the incidence of MD (von Bülow, 1980; Sharma, 1984). Similarly, infection with IBDV during the early latent phase of MDV also caused a decrease in MDV infection levels (Buscaglia *et al.*, 1989). These results are probably caused by the inhibition of T_H cell function as mentioned above, thus preventing or reducing the establishment of MDV infection in T cells.

Reticuloendotheliosis virus

Reticuloendotheliosis virus is clearly an immunosuppressive virus affecting both humoral and cellular immunity (Witter and Fadly, 2003). However, co-infection of REV with MDV did not alter the incidence of MD, nor did REV-infection during the early latent phase of MDV infection decrease MDV viraemia levels (von Bülow 1980; Buscaglia *et al.*, 1989). REV infection did interfere with HVT vaccine efficacy (Witter *et al.*, 1979). Perhaps a more insidious problem with REV is the ability to integrate part or all of its genome into MDV DNA, which is apparently a fairly frequent event under field conditions in certain parts of the world (Davidson and Borenshtain, 2001). The consequence of this type of integration may be that new MDV strains can evolve with increased cytolytic potential, as was shown for the RM1 strain, although this virus was generated by *in vitro* co-culture of MDV and REV.

Reovirus infection

Reoviruses are considered to be immunosuppressive viruses, causing a reduction in lymphocyte mitogenesis (Sharma *et al.*, 1994) probably through the production of NO by activated macrophages (Pertile *et al.*, 1996). There is little information on the importance of reovirus-induced immunosuppression for the pathogenesis of MD. Cho (1979) reported that reovirus infection followed 3 days later by MDV infection caused a reduction in MD incidence. These results are not surprising in view of our current knowledge of the pathogenesis of MD and reovirus-induced immunosuppression. A reduction in the development of activated T cells as a consequence of NO production caused by reovirus infection limits the pool of MDV-susceptible, activated T lymphocytes.

Consequences of immunosuppression

It is obvious that MDV infection causes immunosuppression, but it is not always easy to evaluate the level or relevance of the immunological impairment. *In vitro* assays, such as mitogen stimulation assays, antibody responses to antigens, and specific antigen proliferation assays, have frequently been used to demonstrate MDV-induced immunosuppression (e.g. Rivas and Fabricant, 1988; Reddy *et al.*, 1996), and the relevant literature has recently been reviewed (Schat and Markowski-Grimsrud, 2001; Parcells and Burgess, 2004). Schat and Markowski-Grimsrud (2001) argued that the early immunosuppressive effects as measured by the inhibition of mitogen stimulation of spleen cells or peripheral blood lymphocytes around 7 dpi constitutes a beneficial rather than a negative effect for the host by reducing the pool of MDV-susceptible T cells. This idea is supported by the finding that lymphocytes from MDV-resistant lines often have a lower mitogenic response than lymphocytes from susceptible lines (Calnek *et al.*, 1989b). The mechanism for the decreased mitogen responsiveness during the early cytolytic infection has never been elucidated. Suppressor macrophages have been suggested (Lee *et al.*, 1978), and more recently Schat and Xing (2000) suggested that the activation of iNOS and the subsequent production of NO causes the mitogen unresponsiveness. Increased susceptibility to disease is part of immunosuppression as defined by Dohms and Saif (1984), and the development of MD is often seen as the consequence of MDV-induced immunosuppression. However, apparently healthy chickens can die with massive MD tumours without showing obvious symptoms of immunosuppression. This raises the question: how can we best monitor immunosuppression, and have we selected the best parameters to examine immunosuppression in the context of the pathogenesis of MD?

Although MD is a serious disease in its own right, MDV-induced immunosuppression may aggravate other diseases or reduce vaccine-induced immunity. There are, however, few papers demonstrating an impact of MDV infection on other diseases. MDV infection can increase the severity of challenge with coccidiosis, although this may depend on the genetic resistance of the birds to MDV (Biggs *et al.*, 1968; Rice and Reid, 1973). MDV infection using the vMDV strain HPRS-16 simultaneously with, or followed by, infection with *Cryptosporidium baileyi* prolonged the infection with *C. baileyi*, resulting in lesions associated with this pathogen (Abbassi *et al.*, 1999, 2000). Challenge with MDV caused increased susceptibility to infection with *Escherichia coli* compared to control chickens (Islam *et al.*, 2002). MDV infection reduced the antibody response to *Mycoplasma synoviae* (Kleven *et al.*, 1972; Ellis *et al.*, 1981) and infectious bronchitis virus (IBV) vaccines (Islam *et al.*, 2002).

Neither Newcastle disease nor vaccination against Newcastle disease was influenced by MDV infection using the HPRS-16 strain (Box *et al.*, 1971). Unfortunately, these authors did not show that infection with HPRS-16 caused immunosuppression in their experiments.

MDV vaccines have also been reported to cause mild immunosuppression such as a decrease in antibody production to *Mycoplasma synoviae* (Kleven *et al.*, 1972), and to bovine serum albumin, as well as a temporary increase in susceptibility to *E. coli* (Friedman *et al.*, 1992). These effects were noted with HVT, HVT+SB-1, and CVI988. Others just noted a minor as well as a temporary decrease in circulating lymphocyte numbers after HVT vaccination without affecting antibody titres to IBV vaccine or susceptibility to *E. coli* (Islam *et al.*, 2002). These effects are generally mild and certainly of little consequence for the poultry industry, because MD in non-vaccinated birds is a far greater problem than vaccine-induced immunosuppression.

Summary

MDV infection may interfere with proper immunological functions by several distinct mechanisms all leading to immunosuppression. However, the interactions between virus replication and the immune response are very complex, and the mechanisms causing immunosuppression are not fully understood. Without a doubt additional ways that MDV can cause immunosuppression will be discovered with the advance of microarray techniques and the use of proteomics, leading to a better understanding of the interactions between immune responses and MDV proteins. It will be important that these studies are carried out with different MDV strains in several chicken lines that differ in their resistance to MD so that we gain a better understanding of the continuous evolution of MDV strains and their potential to escape immune responses. With the exception of CIAV, immunosuppressive viruses may be less of a problem in the pathogenesis of MD than has generally been suggested. However, vaccine efficacy may be negatively influenced by these pathogens, especially in commercial poultry operations where multifactorial interactions may have unexpected consequences. Future studies on the pathogenesis of MD must address the question of how immunosuppression can be monitored in a more meaningful way and how MDV-induced immunosuppression can influence other diseases and the general performance of commercial birds.

References

Abbassi, H., Coudert, F., Cherel, Y. *et al.* (1999). *Avian Dis.*, **43**, 738–744.
Abbassi, H., Dambrine, G., Cherel, Y. *et al.* (2000). *Avian Dis.*, **44**, 776–789.
Akaike, T. and Maeda, H. (2000). *Immunology*, **101**, 300–308.

Barrow, A. D., Burgess, S. C., Baigent, S. J. *et al.* (2003). *J. Gen. Virol.*, **84**, 2635–2645.

Biggs, P. M., Long, P. L., Kenzy, S. G. and Rootes, D. G. (1968). *Vet. Rec.*, **83**, 284–289.

Box, P. G., Furminger, I. G. S. and Warden, D. (1971). *Vet. Rec.*, **89**, 475–478.

Bumstead, J. M. and Payne, L. N. (1987). *Vet. Immunol. Immunopathol.*, **16**, 47–66.

Burgess, S. C. and Davison, T. F. (2002). *J. Virol.*, **76**, 7276–7292.

Buscaglia, C., Calnek, B. W. and Schat, K. A. (1989). *Avian Pathol.*, **18**, 265–281.

Bustamante, J., Bersier, G., Romero, M. *et al.* (2000). *Arch. Biochem. Biophysics*, **376**, 239–247.

Calnek, B. W. (1986). *CRC Crit. Rev. Microbiol.*, **12**, 293–320.

Calnek, B. W. (2001). *Curr. Topics Microbiol. Immunol.*, **255**, 25–55.

Calnek, B. W., Schat, K. A., Ross, L. J. *et al.* (1984). *Intl J. Cancer.*, **33**, 389–398.

Calnek, B. W., Lucio, B. and Schat, K. A. (1989a). In *Advances in Marek's Disease Research* (eds S. Kato, T. Horiuchi, T. Mikami and K. Hirai), pp. 324–333. Japanese Association on Marek's Disease, Osaka.

Calnek, B. W., Adene, D. F., Schat, K. A. and Abplanalp, H. (1989b). *Poultry Sci.*, **68**, 17–26.

Calnek, B. W., Carlisle, J. C., Fabricant, J. *et al.* (1979). *Am. J. Vet. Res.*, **40**, 541–548.

Calnek, B. W., Harris, R. W., Buscaglia, C. *et al.* (1998). *Avian Dis.*, **42**, 124–132.

Cho, B. R. (1979). *Avian Dis.*, **23**, 118–126.

Davidson, I. and Borenshtain, R. (2001). *Avian Dis.*, **45**, 102–121.

Derfuss, T. and Meinl, E. (2002). *Curr. Topics Microbiol. Immunol.*, **269**, 257–272.

Djeraba, A., Musset, E., van Rooijen, N. and Quere, P. (2002). *Vet. Microbiol.*, **86**, 229–244.

Dohms, J. E. and Saif, Y. M. (1984). *Avian Dis.*, **28**, 305–310.

Ellis, M. N., Eidson, C. S., Brown, J. *et al.* (1981). *Poultry Sci.*, **60**, 1344–1347.

Ewert, D. and Duhadaway, J. (1999). *Acta Virol.*, **43**, 133–135.

Fabricant, J., Calnek, B. W. and Schat, K. A. (1982). *Avian Dis.*, **26**, 257–264.

Fehler, F. and Winter, C. (2001). In *II International Symposium on Infectious Bursal Disease and Chicken Infectious Anaemia*, pp. 391–394. Institut für Geflugelkrankheiten, Justus Liebig University. Rauischholzhausen, Giessen.

Friedman, A., Shalem-Meilin, E. and Heller, E. D. (1992). *Avian Pathol.*, **21**, 621–631.

Gross, W. B. (1972). *Am. J. Vet. Res.*, **33**, 2275–2279.

Heller, E. D., Levy, A. M., Vaiman, R. and Schwartsburd, B. (1997). *Vet. Immunol. Immunopathol.*, **57**, 289–303.

Holland, M. S., Silva, R. F., Mackenzie, C. D. *et al.* (1994). *Avian Dis.*, **38**, 446–453.

Hunt, H. D., Lupiani, B., Miller, M. M. *et al.* (2001). *Virology*, **282**, 198–205.

Islam, A. F. M. F., Wong, C. W., Walkden-Brown, S. K. *et al.* (2002). *Avian Pathol.*, **31**, 449–461.

Jakowski, R. M., Fredrickson, T. N., Chomiak, T. W. and Luginbuhl, R. E. (1970). *Avian Dis.*, **14**, 374–385.

Jarosinski, K. W., Yunis, R. W., O'Connell, P. H. *et al.* (2002). *Avian Dis.*, **46**, 636–649.

Jarosinski, K. W., O'Connell, P. H. and Schat, K. A. (2003). *Virus Genes*, **26**, 255–269.

Jen, L. W. and Cho, B. R. (1980). *Avian Dis.*, **24**, 896–907.

Jeurissen, S. H. and de Boer, G. F. (1993). *Vet. Q.*, **15**, 81–84.

Jones, D., Brunovskis, P., Witter, R. and Kung, H. J. (1996). *J. Virol.*, **70**, 2460–2467.

Kaiser, P., Underwood, G. and Davison, F. (2003). *J. Virol.*, **77**, 762–768.

Kent, J., Bernberg, E. and Morgan, R. (2001). In *Current Progress on Marek's Disease Research* (eds K. A. Schat, R. M. Morgan, M. S. Parcells and J. L. Spencer), pp. 163–166. American Association of Avian Pathologists, Kennett Square, Pennsylvania.

Kingham, B. F., Zelnik, V., Kopacek, J. *et al.* (2001). *J. Gen. Virol.*, **82**, 1123–1135.

Kleven, S. H., Eidson, C. S., Anderson, D. P. and Fletcher, O. J. (1972). *Am. J. Vet. Res.*, **33**, 2037–2042.

Koszinowski, U. H. and Hengel, H. (eds) (2002). In *Viral Proteins Counteracting Host Defenses*, pp. 1–318. Springer-Verlag, Berlin.

Lee, L. F., Sharma, J. M., Nazerian, K. and Witter, R. L. (1978). *J. Immunol.*, **120**, 1554–1559.

Levy, A. M., Davidson, I., Burgess, S. C. and Dan Heller, E. (2003). *Comp. Immunol. Microbiol. Infect. Dis.*, **26**, 189–198.

Liu, H. C., Kung, H. J., Fulton, J. E. *et al.* (2001). *Proc. Natl Acad. Sci. USA*, **98**, 9203–9208.

Markowski-Grimsrud, C. J. and Schat, K. A. (2003). *Immunology*, **109**, 283–294.

Miles, A. M., Reddy, S. M. and Morgan, R. W. (2001). *Avian Dis.*, **45**, 9–18.

Morgan, R. W., Sofer, L., Anderson, A. S. *et al.* (2001). *J. Virol.*, **75**, 533–539.

Morimura, T., Hattori, M., Ohashi, K. *et al.* (1995). *J. Gen. Virol.*, **76**, 2979–2985.

Morimura, T., Ohashi, K., Kon, Y. *et al.* (1996). *Arch. Virol.*, **141**, 2243–2249.

O'Brien, V. (1998). *J. Gen. Virol.*, **79**, 1833–1845.

Ohashi, K., Mikami, T., Kodama, H. and Izawa, H. (1987). *Intl J. Cancer*, **40**, 378–382.

Parcells, M. S. and Burgess, S. C. (2004). In *Comparative Aspects of Tumor Development*, Vol. 5 (ed. H. E. Kaiser). Kluwer Academic Publishers, Dordrecht (in press).

Parcells, M. S., Anderson, A. S. and Morgan, R. W. (1995). *J. Virol.*, **69**, 7888–7898.

Parcells, M. S., Arumugaswami, V., Prigge, J. T. *et al.* (2003). *Poultry Sci.*, **82**, 893–898.

Payne, L. N., Frazier, J. A. and Powell, P. C. (1976). *Intl Rev. Exp. Pathol.*, **16**, 59–154.

Pertile, T., Karaca, K., Walser, M. M. and Sharma, J. M. (1996). *Vet. Immunol. Immunopathol.*, **53**, 129–145.

Quere, P. (1992). *Vet. Immunol. Immunopathol.*, **32**, 149–164.

Reddy, S. K., Suresh, M., Karaca, K. *et al.* (1996). *Vaccine*, **14**, 1695–1702.

Reddy, S. M., Lupiani, B., Gimeno, I. M. *et al.* (2002). *Proc. Natl Acad. Sci. USA*, **99**, 7054–7059.

Rice, J. T. and Reid, W. M. (1973). *Avian Dis.*, **17**, 66–71.

Rivas, A. L. and Fabricant, J. (1988). *Avian Dis.*, **32**, 1–8.

Roizman, B. (1996). In *Fields Virology*, Vol. 2 (eds D. M. Knipe and P. M. Howley), pp. 221–2230. Lippincott-Raven, Philadelphia.

Schat, K. A. (1987a). *Cancer Surv.*, **6**, 1–37.

Schat, K. A. (1987b). In *Avian Immunology: Basis and Practice* (eds A. Toivanen and P. Toivanen), pp. 101–128. CRC Press, Boca Raton.

Schat, K. A. (2003). In *Diseases of Poultry* (eds Y. M. Saif, H. J. Barnes, A. M. Fadly *et al.*), pp. 182 202. Iowa State Press, Ames.

Schat, K. A. and Calnek, B. W. (1978). *J. Natl. Cancer Inst.*, **60**, 1075–1082.

Schat, K. A. and Markowski-Grimsrud, C. J. (2001). *Curr. Topics Microbiol. Immunol.*, **255**, 91–120.

Schat, K. A. and Xing, Z. (2000). *Dev. Comp. Immunol.*, **24**, 201–221.

Schat, K. A., Calnek, B. W., Chen, C.-L. H. and Benedict, A. A. (1983). In *Leukemia Reviews International*, Vol. 1 (ed. D. S. Yohn), pp. 133–134. Marcel Dekker, Inc., New York.

Schat, K. A., Calnek, B. W., Fabricant, J. and Graham, D. L. (1985). *Avian Pathol.*, **14**, 127–146.

Sharma, J. M. (1984). *Avian Dis.*, **28**, 629–640.

Sharma, J. M., Karaca, K. and Pertile, T. (1994). *Poultry Sci.*, **73**, 1082–1086.

Sharma, J. M., Kim, I.-J., Rautenschlein, S. and Yeh, H. Y. (2000). *Dev. Comp. Immunol.*, **24**, 223–235.

Shaw, I., Ross, N., Venugopal, K. *et al.* (2001). In *Current Progress on Marek's Disease Research* (eds K. A. Schat, R. M. Morgan, M. S. Parcells and J. L. Spencer.), pp. 167–174. American Association of Avian Pathologists, Kennett Square, Pennsylvania.

Shek, W. R., Calnek, B. W., Schat, K. A. and Chen, C.-L. H. (1983). *J. Natl Cancer Inst.*, **70**, 485–491.

Theis, G. A. (1977). *J. Immunol.*, **118**, 887–894.

Theis, G. A. (1981). *Infect. Immun.*, **34**, 526–534.

von Bülow, V. (1980). *Avian Path.*, **9**, 109–119.

Witter, R. L. and Fadly, A. M. (2003). In *Diseases of Poultry* (eds Y. M. Saif, H. J. Barnes, J. R. Glisson *et al.*), pp. 517–536. Iowa State University Press, Ames.

Witter, R. L. and Schat, K. A. (2003). In *Diseases of Poultry* (eds Y. M. Saif, H. J. Barnes, J. R. Glisson *et al.*), pp. 407–464. Iowa State University Press, Ames.

Witter, R. L., Lee, L. F., Bacon, L. D. and Smith, E. J. (1979). *Infect. Immun.*, **26**, 90–98.

Witter, R. L., Li, D., Jones, D. *et al.* (1997). *Avian Dis.*, **41**, 407–421.

Xing, Z. and Schat, K. A. (2000). *Immunology*, **100**, 70–76.

Zanella, A. and Dall'Ara, P. (2001). In *Current Progress on Marek's Disease Research* (eds K. A. Schat, R. M. Morgan, M. S. Parcells and J. L. Spencer.), pp. 11–19. American Association of Avian Pathologists, Kennet Square, Pennsylvania.

Diagnosis of Marek's disease

12

VLADIMÍR ZELNÍK
Lohmann Animal Health GmbH & Co., Cuxhaven, Germany

Introduction

Intensive and widespread vaccination of nearly all commercial poultry flocks during the last 30 years has significantly reduced losses from Marek's disease (MD) around the world. However, sporadic outbreaks still occur in many commercial flocks. For this reason, accurate and rapid methods are crucial for diagnosing MD and identifying the causative, pathogenic isolates of serotype 1 Marek's disease virus (MDV-1). Diagnosis of MD in the field is mostly based on clinical signs and examination of gross or microscopic lesions in tissues. MDV infections that do not induce overt pathological changes can be confirmed by laboratory-based virological techniques such as virus isolation, identification of MDV-specific antigens or antibodies, as well as viral nucleic acids. It is important to recognize that pathological diagnostic techniques identify the nature of the tumour that is causing mortality, whereas virological diagnosis identifies viruses that are present in the bird or the flock. Because MDV commonly occurs in poultry flocks, virological diagnosis alone does not necessarily establish the cause of the tumour. Nevertheless, pathological confirmation of the disease together with the positive identification of MDV often confirms the existence of MD in a flock.

Viruses of all three MDV serotypes: attenuated strains of MDV-1 as well as the non-pathogenic serotype 2 and herpesvirus of turkeys (HVT: serotype 3 MDV), are used as live vaccines against MD. Although these vaccines protect birds against the disease, they do not protect them from superinfection with pathogenic MDV. Consequently, MDV can persist in the birds throughout their lives, and can spread horizontally between the birds in a flock following infection from virus particles released from feather epithelium as dust and dander. In this form, MDV can remain infectious and relatively resistant to environmental conditions for long periods. Intensive poultry production methods and the wide-spread use of vaccines are the two most likely factors contributing to increased virulence of MDV isolates (see Chapters 5 and 14). In addition, some of the more recent hypervirulent (very virulent (vv)MDV and vv+MDV) pathotypes induce non-classical signs of disease, such as a dramatic early cytolytic infection (see Chapter 7).

As an evolving pathogen, MDV requires continuous monitoring with regard to its pathogenesis, as well as the development of new strategies for prevention, in conjunction with rapid diagnosis. This chapter briefly summarizes the examination of gross pathological changes during an MD outbreak. Furthermore, as it is becoming more and more important to confirm preliminary findings in the field with laboratory diagnostic techniques, this chapter will discuss both the classical and new techniques for MDV identification and characterization.

Diagnosis of MD under field conditions

Marek's disease displays a very complex pathology. A major step forward came with Biggs (1961) and Campbell (1961), who proposed differentiating MD aetiologically from other lymphoproliferative diseases and naming the condition after József Marek (1907), who first described it as polyneuritis (see Chapter 2). Typical indications of MDV infection in a flock include paralysis as a result of lymphoid infiltration into peripheral nerves, lymphomas in various organs, immunosuppression, severe depression, blindness, and skin lesions, all of which can be accompanied by non-specific signs such as weight loss, anorexia and pallor. Currently, MD symptoms are classified as the 'classical' neural form of fowl paralysis, acute leukosis with lymphomatous tumours in visceral organs, transient paralysis (Payne *et al.*, 1976) and the more recently defined acute transient paralysis (Witter *et al.*, 1999). The division between classical forms, acute forms and transient paralysis is usually not very distinct, as similar symptoms can be associated with all forms of MD. All these pathological changes are associated with infiltration of proliferating lymphoid cells in organs such as liver, kidney, spleen, gonads, heart, proventriculus, skeletal muscle, skin and peripheral nerves. However, other clinical signs and early mortality associated with severe cytolytic infection of lymphoid tissues can also be observed in birds infected by vvMDV or vv+MDV pathotypes. More detailed information on the pathological changes associated with MDV infection can be found in Chapter 7 as well as in other reviews (Payne, 1985; Sharma, 1985; Calnek, 2001).

The pathogenesis of MDV in susceptible birds can be greatly influenced by other immunosuppressive viral or bacterial infections (see Chapter 11). Of particular interest, and most specifically linked with many MD outbreaks, is infection with chicken infectious anaemia virus (CIAV; Otaki *et al.*, 1988; Miles *et al.*, 2001). Co-infection of birds with avian leukosis virus (ALV) and MDV

Table 12.1 Useful features for differentiating Marek's disease from lymphoid leukosis.

Feature	Marek's disease	Lymphoid leukosis
Age	6 weeks and more	Not less than 16 weeks
Symptoms	Frequent paralysis	Non-specific
Incidence in flocks	Frequently above 5% in unvaccinated flocks	Rarely above 5%
Macroscopic lesions:		
• Neural enlargements	Frequent	Absent
• Bursa of Fabricius	Diffuse enlargement or atrophy	Nodular tumours
• Tumours in skin, muscle and proventriculus	May be present	Usually absent
Microscopic lesions:		
• Neural involvement	Yes	No
• Liver tumours	Often perivascular	Focal or diffuse
• Spleen	Diffuse	Often focal
• Central nervous system	Yes	No
• Lymphoid proliferation in skin	Yes	No
Cytology of tumours	Pleomorphic cells including lymphoblasts, small, medium and large lymphocytes and reticulum cells	Lymphoblasts
Category of neoplastic lymphoid cells	T cells	B cells

can also occur, resulting in depression of MD vaccine-induced immunity (Witter *et al.*, 1979) and potential integration of retroviral sequences into the MDV genome (Isfort *et al.*, 1992; Davidson *et al.*, 1995). Single or potential dual infections of chickens with ALV and/or MDV also have important consequences for the correct differential diagnosis of MD or lymphoid leukosis (LL) due to the similar pathology induced by these viruses. Some of the useful features enabling differentiation of MD from LL (see Biggs, 1976; Powell and Payne, 1993) are provided in Table 12.1.

It is clear that, together with clinical signs, additional factors indicating an outbreak of MD in a flock have to be monitored. These mainly include: the vaccination programme (properly prepared vaccine, dose administered and vaccine type), overall flock management, the age of affected birds, and the occurrence of other, mainly immunosuppressive, infections that can dramatically affect MD outbreaks (see Chapter 5). In general, good laboratory diagnostic practices should be used following the first suspicion of an outbreak of MD in a flock. One of the most reliable methods for verification is to inoculate susceptible chickens with the infectious agent and confirm MD by clinical signs in the recipients.

Laboratory diagnosis of MD

Laboratory diagnosis of MD mostly concerns isolation of the virus, followed by identification and characterization of its virions, its DNA or antigens, as well as detection of induced MDV-specific antibodies. It should be mentioned that although convalescent MD sera are available commercially for some of the traditional MDV detection tests, commercial testing kits cannot be obtained and therefore diagnostic laboratories have to rely on setting up their own diagnostic techniques. However, recent progress with immunological and molecular biology techniques has enabled more rapid, sensitive and accurate detection of MDV in biological samples from routine MD investigations. For reliable laboratory diagnosis, there are recommended procedures. It is always best to include two independent techniques for MDV detection, such as virus isolation and very sensitive PCR analysis. MD vaccine viruses are also likely to be detected, and MDV-1 vaccines especially can confuse positive MD diagnosis. Differentiation between MD vaccine viruses and field isolates is discussed below.

Collection of infectious material for *in vitro* studies

One of the important requirements for reliable diagnosis of MD, unfortunately often overlooked, is proper collection of infectious material. Especially when attempting virus isolation *in vitro*, it is essential to process samples from affected birds as soon as possible. MDV is a highly cell-associated virus, and therefore it is essential that the test material consists of viable cells. Non-coagulated whole blood, collected in heparin or citrate buffer, or isolated peripheral white blood cells can be used as the primary inoculum for *in vitro* isolation of MDV and/or for infecting susceptible birds. In the same way, single cell suspensions prepared from lymphoid organs such as the spleen or from tumours can serve as the primary inoculum. Samples of cell suspensions should be treated like live, cell-associated virus, and stored in liquid nitrogen ($-196°C$) in the presence of serum and cryoprotectants such as dimethylsulphoxide. These biological samples can also be used for identification of MDV antigens or DNA. However, for this purpose, cryopreservation of the samples is not necessary. The collected tissue samples can be stored at $4°C$ for a few days, or better still at $-20°C$, pending subsequent isolation and detection of viral DNA. For immunohistological analyses it is advisable to fix the tissues appropriately. The feather follicles from infected birds also provide a good source of MDV antigens and viral DNA, and can be used (Calnek *et al.*, 1970; Haider *et al.*, 1970; Davidson *et al.*, 1986). Tips of feathers from affected birds can be stored in a cold, dry place for several weeks pending antigen or DNA analyses.

Virus isolation and identification

Electron microscopic examination of infected material was mainly used in early studies that analysed the morphology and structure of MDV (Nazerian and Burmester, 1968; Nazerian *et al.*, 1971; Hirumi *et al.*, 1974). Because of the need for specialist and expensive equipment, it is not routinely used for identifying MDV. MDV differs from related α-herpesviruses in many of its biological properties, so for studies on virus morphology, envelopment and spread during infection, modern electron microscopic techniques have a significant role to play.

In vitro isolation and propagation of the causative agent is a prerequisite for further characterization of the MDV isolate. However, this is time consuming, and requires both skill and good experimental facilities for the successful adaptation of field isolates to *in vitro* culture. Although chicken embryo fibroblasts (CEF) are common cells for the production of MD vaccines, they are not always suitable for primary isolation of MDV. Chicken kidney cells (CKC) or duck embryo fibroblasts (DEF) seem to be much more suitable cells when it comes to primary propagation of MDV isolates (Churchill and Biggs, 1967; Nazerian *et al.*, 1968; Solomon *et al.*, 1968). Virus adaptation to *in vitro* cell culture is usually successful after co-cultivation of the primary inoculum with DEF or CKC. The number of cells in the inoculum is also very important for successful isolation. However, since this can be highly variable and is not predictable, it is better to seed different numbers of cells in replicate plates of reporter cells to maximize the likelihood of virus isolation. For isolation of attenuated strains of MDV-1 or the other serotypes from biological samples, CEF can provide an alternative substrate. Although there have been reports of successful propagation of MDV in some continuous cell lines, such as OU2.2, OU2.1 (Abujoub and Coussens, 1995), QM7 (Schumacher *et al.*, 2000) and SOgE (Schumacher *et al.*, 2002), their use for the primary isolation of MDV from infected samples has not been described.

In general, virulent field isolates of MDV propagate only slowly in cell culture, with the cytopathic effect (CPE) comprising small to medium-sized plaques of rounded refractile cells or syncytia (Plate 7) appearing 5–8 days post-inoculation (dpi). Continuous passage of the virus in cell culture leads to its adaptation and faster replication *in vitro*. However, this is associated with attenuation of the oncogenic potential of the virus (Churchill *et al.*, 1969; Nazerian, 1970) and structural and gene expression changes (Churchill *et al.*, 1969; Fukuchi *et al.*, 1985; Maotani *et al.*, 1986; Ross *et al.*, 1993; Wilson *et al.*, 1994). Attenuated MDV strains produce larger plaques and the CPE usually occurs 2–4 dpi (Plate 8). On the basis of plaque size and morphology, a tentative identification of the isolate can be made, although further analysis needs to be carried out for the precise characterization of the virus. For example, the infected cell culture could be examined by immunohistochemical or immunofluorescent techniques using antibodies recognizing MDV antigens (Plates 7B, 8B) for a more specific identification.

Detection and characterization of MDV antigens and antibodies

Immunological methods for identifying MDV antigens or detecting MDV-specific antibodies in sera from infected birds were generally used after successful virus isolation in the 1960s. One such method, which is still in use, is the agar gel precipitation (AGP) assay where serum is reacted with MDV antigen (Chubb and Churchill, 1968). A crude lysate from *in vitro* infected cells, skin extracts or feather pulps can serve as the source of MDV antigen (Haider *et al.*, 1970; Adldinger and Calnek, 1973). The presence of either MDV antigen or MDV-specific sera can be detected using this assay. It is a simple test, and the antigen suspension or even tips obtained directly from small feathers are placed in wells prepared in an agar layer containing salt and buffer. Serum is placed in adjacent wells and the samples are left to diffuse and react for 24–72 hours in a humidified atmosphere; a positive reaction can be identified by the precipitates in the agar between the wells. In the precipitate, the MDV antigen A predominates; this was later identified as a homologue of herpesvirus glycoprotein C (gC; Binns and Ross, 1989). It should be mentioned, however, that expression of gC becomes significantly reduced in highly-passaged, attenuated strains of MDV (Churchill *et al.*, 1969; Ikuta *et al.*, 1983a). Lower levels of virus isolated from lymphocytes (Schat *et al.*, 1985) and the inability to detect MDV antigens by immunofluorescence in the feather-follicle epithelium (Maas *et al.*, 1978) suggest that attenuated MDV viruses have reduced replication *in vivo*. These findings suggest that the AGP test could provide a simple means of differentiation between virulent and attenuated/vaccine MDV strains.

Direct or indirect immunofluorescent assays can also be used as methods for detecting MDV-specific antibodies or antigens in samples. For the detection of MDV-specific antibodies in serum samples, chicken antibodies directly conjugated to a fluorescent tag, or chicken antibodies reacted with fluorescent-conjugated secondary anti-chicken antibody, are examined for positive fluorescence signals using a fixed MDV-infected tissue culture monolayer. Utilization of secondary-labelled anti-chicken antibodies usually provides a stronger and more specific signal. In the reverse format, convalescent MD serum, monospecific or monoclonal antibodies can be used to detect the presence of MDV antigens in various infected samples either *ex vivo* and/or *in vitro*. The latter system is more useful as it enables more specific detection of MDV antigens in tissues or infected cells, allowing their further characterization, especially if monospecific sera or better still monoclonal antibodies are available. Availability and use of monoclonal antibodies has opened up a new era for detailed analyses of MDV and HVT antigens and detection and differentiation of the viruses (Ikuta *et al.*, 1982; Lee *et al.*, 1983b; Dorange *et al.*, 2000). Preparation of polyclonal antibodies directed against MDV-specific proteins, such as pp38 or MEQ, provides an alternative means for the sensitive and specific detection of virus (see Plates 7B, 8B). Monoclonal antibodies or monospecific sera can be used in various immunological techniques; however, for diagnostic purposes, immuno-fluorescence (Ikuta *et al.*, 1982; Lee *et al.*, 1983b) or immunoperoxidase staining (Silva *et al.*, 1997) of *in vitro*-propagated virus isolates appear to be the most useful and straightforward.

Polyclonal or monoclonal antibodies that specifically react with the so-called Marek's disease tumour-associated surface antigen (MATSA) have been prepared by immunizing with *in vitro*-cultured MDV-transformed lympho-blastoid cells (Witter *et al.*, 1975; Lee *et al.*, 1983a). These antibodies have been used extensively for diagnostic immunostaining of materials from infected birds. However, MATSA is a host cell antigen that in some cases is also a marker on activated T cells, and not all MDV-transformed cells express it. One of the MATSA antigens is CD30 (see Chapter 8), recognized by the mono-clonal antibody AV37, and overexpressed in MDV-transformed cells (Burgess and Davison, 2002). Significantly higher proportions of AV37-positive cells can also be detected in peripheral blood leukocytes using flow cytometry early after MDV infection (Burgess and Davison, 2002). For more details on the nature and expression of AV37 antigen, the reader is referred to Chapter 8.

An enzyme-linked immunosorbent assay (ELISA) has also been developed for the detection of MDV-specific antibodies (Cheng *et al.*, 1984), although it has not found widespread diagnostic use. There are two pertinent reasons for this. First, MD vaccines contain suspensions or extracts from the live-infected CEF on which they have been cultured, and these can induce antibodies against antigens expressed by CEF. Hence this often results in high background readings in ELISA, especially if the coating MDV antigens used in the ELISA have been prepared using CEF. Secondly, MDV is an immunosuppressive virus that depletes the antibody-producing B cells and causes a reduction in titres of MDV-specific antibodies. In our experience, when a different cell culture system is used for the production of the MDV coating antigen (e.g. CKC), ELISA readings are more reliable (Zelnik *et al.*, 2004). Such assays can be used to distinguish between birds that have been infected and are protected against MDV from those that received vaccine virus only, or were not protected against MDV.

Identification of MDV DNA

Traditional techniques for specific DNA detection by hybridization with MDV-specific probes have not found much use in MD diagnosis, as they are both labori-ous and time-consuming. However, introduction of the polymerase chain reaction (PCR) in the 1980s has revolutionized the rapid detection of many infectious agents, including MDV. The main advantages of the PCR method are that it is rapid, and DNA isolated even from crude biological materials can be analysed. As the complete DNA sequences of the MDV-1, MDV-2 and HVT genomes are available (Tulman *et al.*, 2000; Afonso *et al.*, 2001; Izumiya *et al.*, 2001; Kingham *et al.*, 2001), manual or software-assisted design of specific PCR primers is not a problem and just depends on actual requirements and purpose.

In this section, the important steps for setting up successful and reliable molecular diagnosis of MDV will be summarized. Specific PCR applications for differentiation of viruses will be discussed in the following section.

First, it should be mentioned that PCR is a very sensitive system and every precaution must be taken to avoid cross-contamination. Good laboratory practice is essential at all stages in the process, starting with the collection of materials, the isolation of DNA, and finally setting up the PCR itself. Choice of primers usually depends on the users and their requirements. However, for designing efficient diagnostic PCR assays it is advisable to use computer programs for primer design, so that suitable primers can be identified with respect to the DNA sequence to be amplified. Potential cross-reactivity with related viruses can be easily investigated. In general, synthesis of primers with a melting temperature of more than 55°C is recommended to avoid non-specific annealing and synthesis. The PCR has to be optimized for each primer set, depending on the thermocycler used, as well as the various reagents such as the buffer and heat-stable DNA polymerase. Although the calculated annealing temperatures can serve as a guide, it is recommended that the optimal annealing temperature be determined experimentally, as this can be affected by the different buffers and enzyme formulations provided by different suppliers. Use of 'hot-start' DNA polymerase formulations is recommended. Optimal concentration of Mg^{2+} ions is also one of the essential factors influencing a successful PCR. For every set of samples, positive and negative controls must be included. Further references for setting up the PCR can be found in the numerous PCR manuals available. Recently, the use of quantitative PCR technique has become more popular; however, because specialized, expensive equipment and consumables are required, its general use in routine MDV detection and diagnosis is not a practical proposition.

Differentiation of MDV serotypes, strains and isolates

Since MDV is ubiquitous, the use of vaccines is widespread and the detection methods are becoming more sensitive, differentiation between virulent and vaccine viruses has become much more important. Use of MDV-1 vaccines, such as CVI988/Rispens (Rispens *et al.*, 1972), has further underlined this need. Some of the possible methods of virus differentiation were outlined in earlier sections; however, many of them, such as virus plaque morphology, are not specific and require skill and experience. A simple method for discriminating HVT from MDV is based on the ability of the isolate to be propagated in the QT35 cell line (Cho, 1981). However, recent studies have shown that QT35 cells contain a latent MDV genome that can be transactivated by co-infection with HVT (Yamaguchi *et al.*, 2000) or adapted by the CVI988/Rispens vaccine virus (Majerciak *et al.*, 2001). It is clear that reliable methods for differentiating viruses need to be based on specific and sensitive antigen or DNA detection methods.

MDV-1, MDV-2 and HVT are serologically related viruses possessing cross-reactive antigenic determinants (Ikuta *et al.*, 1983b), and the use of convalescent sera for differentiation gives non-specific results. On the other hand, monoclonal antibodies provide an excellent means of virus discrimination, as they recognize specific epitopes on antigens. Most of these monoclonal antibodies are serotype-specific, although a few display some cross-reactivity. One of these monoclonal antibodies, H19, that recognizes the MDV-encoded pp38 antigen (Lee *et al.*, 1983b), can differentiate between CVI988/Rispens vaccine virus and other MDV isolates; the CVI988 virus contains mutations that have altered the pp38 antigenic epitope, making it non-reactive with the H19 monoclonal antibody (Cui *et al.*, 1999). This is the only example of immunological discrimination between serotype 1 virus isolates. Polyclonal antibodies against virus-specific proteins, such as pp38, can be also used to differentiate between serotypes; however, their use to discriminate between different isolates within a serotype is usually not possible.

With correct choice of primer sets for PCR, it is easy to establish PCR method(s) for differentiating between viruses belonging to different MDV serotypes. On the other hand, as there is only limited information on genetic variability between the viruses within a given serotype, PCR differentiation between such MDV isolates is much more difficult. Visible electrophoretic changes between virulent MDV isolates and their attenuated counterparts were evident in the *Bam*HI-H fragment of the MDV genome, as an expansion of this region (Fukuchi *et al.*, 1985; Maotani *et al.*, 1986). The expansion is due to multiplication of 132 base pair (bp) repeats that are present in virulent strains as only two or three copies, whereas in highly-passaged viruses their number is variable and can reach several tens (Ross *et al.*, 1993). This characteristic has enabled the

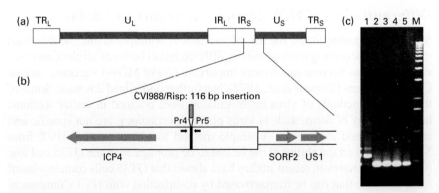

Figure 12.1 Location and PCR analysis of MDV, CVI988/Rispens strain-specific insertion. (a) Genomic map of MDV; (b) junction of Us/IRs genomic elements with location of the insertion; (c) scanned image of an ethidium bromide-stained agarose gel using PCR products from the following MDV strains[1] DNA templates, CVI988/Rispens (1), HPRS16 low passage (2), HPRS16 high passage (3), RB1B (4), EU1 (5).

development of a simple PCR test to differentiate between virulent and passaged viruses (Silva, 1992; Becker *et al.*, 1993). PCR analysis of the expansion of the 132 bp repeats during multiple *in vitro* passage seems to be a reliable method for identifying attenuated viruses, including the CVI988/Rispens strain. However, it is important to bear in mind that expansion of the 132 bp repeats is only a marker of attenuation and not the cause (Silva *et al.*, 2004; V. Zelnik, unpublished data). Variability in the MDV-specific *meq* gene has also been reported, but it appears that this polymorphism cannot be used for differentiating vaccine and virulent viruses (Chang *et al.*, 2002). Another CVI988/Rispens-specific insertion is represented by a single 116 bp repeat upstream of the MDV ICP4 ORF (Majerciak *et al.*, 2001), and can be easily detected by PCR (Figure 12.1). As in the case of 132 bp repeat expansion in the *Bam*HI-H region, the nature of this single repeat in the region upstream of the ICP4 open reading frame is not clear, and most likely can be attributed to virus passage in heterologous cell systems.

Summary

MD continues to be a serious threat to poultry production, despite widespread use of vaccination programmes. Sporadic outbreaks of MD continue to occur and have been associated with increased virulence in field strains of MDV, the presence of other (mainly immunosuppressive) infectious agents, and poor flock management. Rapid and reliable diagnosis of MD remains an important issue. Alongside classical methods, such as examination of gross pathology, novel laboratory diagnostic methods have gained growing significance. Individual laboratories have to rely on developing their own reliable and reproducible methods, as no standard methodologies are available. This chapter has attempted to provide an overview and some guidance on the steps to take following a suspected MD outbreak in a flock, and to outline recently developed immunological and molecular laboratory diagnostic methods.

References

Abujoub, A. and Coussens, P. M. (1995). *Virology*, **214**, 541–549.
Adldinger, H. K. and Calnek, B. W. (1973). *J. Natl Cancer Inst.*, **50**, 1287–1298.
Afonso, C. L., Tulman, E. R., Lu, Z. *et al.* (2001). *J. Virol.*, **75**, 971–978.
Becker, Y., Tabor, E., Asher, Y. *et al.* (1993). *Virus Genes*, **7**, 277–287.
Biggs, P. M. (1961). *Br. Vet. J.*, **117**, 326–334.
Biggs, P. M. (1976). In *Differential Diagnosis of Avian Lymphoid Leukosis and Marek's Disease* (ed. L. N. Payne), pp. 67–85. Commission of the European Communities, Luxembourg.
Binns, M. M. and Ross, L. J. N. (1989). *Virus Res.*, **12**, 371–382.
Burgess, S. C. and Davison, T. F. (2002). *J. Virol.*, **76**, 7276–7292.
Calnek, B. W. (2001). In *Marek's Disease* (ed. K. Hirai), pp. 25–55. Springer-Verlag, Berlin.

Calnek, B. W., Adlinger, H. K. and Kahn, D. E. (1970). *Avian Dis.*, **14**, 219–233.

Campbell, J. G. (1961). *Br. Vet. J.*, **117**, 316–325.

Chang, K.-S., Ohashi, K. and Onuma, M. (2002). *J. Vet. Med. Sci.*, **64**, 1097–1101.

Cheng, Y.-Q., Lee, L. F., Smith, E. J. and Witter, R. L. (1984). *Avian Dis.*, **28**, 900–911.

Cho, B. R. (1981). *Avian Dis.*, **25**, 839–846.

Chubb, R. C. and Churchill, A. E. (1968). *Vet. Rec.*, **83**, 4–7.

Churchill, A. E. and Biggs, P. M. (1967). *Nature*, **215**, 528–530.

Churchill, A. E., Chubb, R. C. and Baxendale, W. (1969). *J. Gen. Virol.*, **4**, 557–564.

Cui, Z., Qin, A., Lee, L. F. *et al.* (1999). *Acta Virol.*, **43**, 169–173.

Davidson, I., Maray, T., Malkinson, M. and Becker, Y. (1986). *J. Virol. Meth.*, **13**, 231–244.

Davidson, I., Borowsky, A., Perl, S. and Malkinson, M. (1995). *Avian Pathol.*, **24**, 69–94.

Dorange, F., El Mehdaoui, S., Pichon, C. *et al.* (2000). *J. Gen. Virol.*, **81**, 2219–2230.

Fukuchi, K., Tanaka, A., Schierman, L. W. *et al.* (1985). *Proc. Natl Acad. Sci. USA*, **82**, 751–754.

Haider, S. A., Lapen, R. F. and Kenzy, S. G. (1970). *Poultry Sci.*, **49**, 1654–1657.

Hirumi, H., Frankel, J. W., Pricket, C. O. and Maramorosch, K. (1974). *J. Natl Cancer. Res.*, **52**, 303–306.

Ikuta, K., Honma, H., Maotani, K. *et al.* (1982). *Biken J.*, **25**, 171–175.

Ikuta, K., Ueda, S., Kato, S. and Hirai, K. (1983a). *J. Gen. Virol.*, **64**, 2597–2610.

Ikuta, K., Ueda, S., Kato, S. and Hirai, K. (1983b). *J. Gen. Virol.*, **64**, 961–965.

Isfort, R., Jones, D., Kost, R. *et al.* (1992). *Proc. Natl Acad. Sci. USA*, **89**, 991–995.

Izumiya, Y., Jang, H. K., Ono, M. and Mikami, T. (2001). In *Marek's Disease* (ed. K. Hirai), pp. 191–221. Springer-Verlag, Berlin.

Kingham, B. F., Zelnik, V., Kopacek, J. *et al.* (2001). *J. Gen. Virol.*, **82**, 1123–1135.

Lee, L. F., Liu, X., Sharma, J. M. *et al.* (1983a). *J Immunol.*, **130**, 1007–1011.

Lee, L. F., Liu, X. and Witter, R. L. (1983b). *J Immunol.*, **130**, 1003–1006.

Maas, H. J. L., van Vloten, J., Vreede-Groenewegen, A. E. and Orthel, F. W. (1978). *Avian Pathol.*, **7**, 79–86.

Majerciak, V., Valkova, A., Szabova, D. *et al.* (2001). *Acta Virol.*, **45**, 101–108.

Maotani, K., Kanamori, A., Ikuta, K. *et al.* (1986). *J. Virol.*, **58**, 657–660.

Marek, J. (1907). *Dtsch Tierärztl. Wochenschr.*, **15**, 417–421.

Miles, A. M., Reddy, S. M. and Morgan, R. W. (2001). *Avian Dis.*, **45**, 9–18.

Nazerian, K. (1970). *J. Natl Cancer Inst.*, **44**, 1256–1267.

Nazerian, K. and Burmester, B. R. (1968). *Cancer Res.*, **28**, 2454–2462.

Nazerian, K., Solomon, J. J., Witter, R. L. and Burmester, B. R. (1968). *Proc. Soc. Exp. Biol. Med.*, **127**, 177–182.

Nazerian, K., Lee, L. F., Witter, R. L. and Burmester, B. R. (1971). *Virology*, **43**, 422–452.

Otaki, Y., Nunoya, T., Tajima, M. *et al.* (1988). *Avian Pathol.*, **17**, 333–347.

Payne, L. N. (1985). In *Marek's Disease* (ed. L. N. Payne), pp. 43–75. Martinus Nijhoff, Boston.

Payne, L. N., Frazier, J. A. and Powell, P. C. (1976). *Int. Rev. Exp. Path.*, **16**, 59–154.

Powell, P. C. and Payne, L. N. (1993). In *Virus Infections of Birds* (eds. J. B. McFerran and M. S. McNulty), pp. 66–70. Elsevier Science Publishers, Amsterdam.

Rispens, B. H., van Vloten, H., Mastenbroek, N. *et al.* (1972). *Avian Dis.*, **16**, 108–125.

Ross, N., Binns, N. M., Sanderson, M. and Schat, K. A. (1993). *Virus Genes*, **7**, 33–51.

Schat, K. A., Calnek, B. W., Fabricant, J. *et al.*, (1985). *Avian Pathol.*, **14**, 127–146.

Schumacher, D., Tischer, B. K., Fuchs, W. and Osterrieder, N. (2000). *J. Virol.*, **74**, 11088–11098.

Schumacher, D., Tischer, B. K., Teifke, J.-P. *et al.* (2002). *J. Gen. Virol.*, **83**, 1987–1992.

Sharma, J. M. (1985). In *Marek's Disease* (ed. L. N. Payne), pp. 151–175. Martinus Nijhoff, Boston.

Silva, R. F. (1992). *Avian Dis.*, **36**, 521–528.

Silva, R. F., Calvert, J. G. and Lee, L. F. (1997). *Avian Dis.*, **41**, 528–534.

Silva, R. F., Reddy, S. M. and Lupiani, B. (2004). *J. Virol.*, **78**, 733–740.

Solomon, J. J., Witter, R. L., Nazerian, K. and Burmester, B. R. (1968). *Proc. Soc. Exp. Biol. Med.*, **127**, 173–177.

Tulman, E. R., Afonso, C. L., Lu, Z. *et al.* (2000). *J. Virol.*, **74**, 7980–7988.

Wilson, W. R., Southwick, R. A., Pulaski, J. T. *et al.* (1994). *Virology*, **199**, 393–402.

Witter, R. L., Stephens, E. A., Sharma, J. M. and Nazerian, K. (1975). *J. Immunol.*, **115**, 177–183.

Witter, R. L., Lee, L. F., Bacon, L. D. and Smith, E. J. (1979). *Infect. Immun.*, **26**, 90–98.

Witter, R. L., Gimeno, I. M., Reed, W. M. and Bacon, L. D. (1999). *Avian Dis.*, **43**, 704–720.

Yamaguchi, T., Kaplan, S. L., Wakenell, P. and Schat, K. A. (2000). *J. Virol.*, **74**, 10176–10186.

Zelnik, V., Harlin, O., Fehler, F. *et al.* (2004). *J. Vet. Med. B. Infect. Dis. Vet. Public Health*, **51**, 61–67.

Vaccination against Marek's disease

13

MICHEL BUBLOT* and JAGDEV SHARMA**

*Virology Unit, Discovery Research, Merial, Lyon, France
**Department of Veterinary and Biomedical Sciences, College of Veterinary Medicine, University of Minnesota, St Paul, Minnesota, USA

History of Marek's disease vaccine

The first vaccine against Marek's disease (MD) was described (Churchill *et al.*, 1969) shortly after identification of the causative agent Marek's disease virus (MDV), a cell-associated herpesvirus (Churchill and Biggs, 1967; Nazerian *et al.*, 1968). This vaccine was based on the oncogenic HPRS-16 strain of serotype 1 MDV (MDV-1) that had been attenuated by serial passages using chicken kidney cell cultures. It was licensed in the UK in 1970, but was soon replaced by a new vaccine, based on herpesvirus of turkey (HVT) (Okazaki *et al.*, 1970; Witter *et al.*, 1970), designated serotype 3 MDV (MDV-3) or *Meleagrid herpesvirus 1* (see Chapter 3). The HVT vaccine (FC126 strain) was initially licensed in the USA in 1971, but was rapidly taken up and used by the poultry industry worldwide. The HVT vaccine was developed as a cell-associated vaccine (a suspension of viable infected cells, also called a 'wet' vaccine), but later it became available as a cell-free lyophilized vaccine ('dry' vaccine). The latter was used mainly in countries where the necessary cold storage for retaining the viability of the wet vaccine was a problem. HVT-based vaccines are still widely used, either alone or in association with other vaccine serotypes.

Soon after the identification of MDV, Rispens *et al.* (1969) reported that inoculation of an MDV-1 isolate of low pathogenicity into day-old MD-susceptible chicks protected them from mortality and gross pathological lesions. An attenuated vaccine strain, designated CVI988 but also called the Rispens strain, was developed and shown to be protective in both laboratory and field trials (Rispens *et al.*, 1972a, 1972b). The Rispens vaccine was licensed in 1973 in The Netherlands, and was later used in other European, as well as Asian, countries.

Some safety problems were reported with highly susceptible birds (von Bülow, 1977), and therefore safer vaccines, designated CVI988 clone C and clone C/R2, were developed from the original vaccine strain and licensed for use in several countries, including the USA (Pol *et al.*, 1986; de Boer *et al.*, 1986, 1987). However, these vaccines were found to be less efficacious than the original CVI988 vaccine strain. The CVI988 vaccine was launched much later (1994) in the USA. In early studies, the Rispens and HVT vaccines provided similar levels of protection (Vielitz and Landgraf, 1971; Maas *et al.*, 1982), but later the Rispens vaccine was shown to provide better protection against highly virulent challenge strains of MDV (Witter *et al.*, 1995). It is now considered to be the 'gold standard' for MD vaccines. Currently, most of the parental flocks (breeders) and layers are vaccinated at 1 day of age with the Rispens vaccine.

In 1978, Schat and Calnek characterized an apparently non-oncogenic strain, SB-1 (Schat and Calnek, 1978), belonging to the MDV serotype 2 (MDV-2). On its own this virus strain was able to protect against pathogenic strains of MDV-1, but when administered with HVT it had synergistic activity, providing improved protection (Calnek *et al.*, 1983; Witter and Lee, 1984). The observation prompted development of a bivalent HVT+SB-1 vaccine that had a higher protection index than HVT. The bivalent vaccine was introduced into the USA market in 1983 and is still widely used, mainly in North America and Japan, although it has not been licensed for the European market. Another MDV-2 strain, 301B/1, was later licensed in the USA (Witter, 1987; Witter *et al.*, 1987). Experimental data suggested that the use of MDV-2 vaccines enhanced levels of lymphoid leukosis (LL) in laying stock (Bacon *et al.*, 1989). This enhancement occurred in the field in some commercial lines, which were quickly replaced by the poultry industry, and is no longer a problem. Witter (1995) developed an attenuated vaccine lacking the ability to enhance LL, but this vaccine was never launched. HVT+MDV-2 bivalent vaccines provide an intermediate level of protection somewhere between that induced by HVT alone and the CVI988 vaccine.

Other MDV-1 vaccine strains have been developed, including the Md11/75C/R2/23 strain, licensed in the USA in 1994 (Witter *et al.*, 1995), and the BH16 vaccine, derived from an Australian very-virulent (vv) MDV isolate (Karpathy *et al.*, 2002, 2003). The former is less protective than the original Rispens vaccine, while the latter is claimed to be at least as protective as the Rispens vaccine (Karpathy *et al.*, 2002, 2003). Additional field trials are necessary to compare the efficacy of the Rispens vaccine with this new vaccine strain.

Adjuvants have been tested to improve the protection provided by MD vaccines. In 1992, Acemannan, an acetylated mannose polymer extracted from the *Aloe vera* plant, was licensed in the USA for use with HVT vaccine. This product has been shown to advance the onset of immunity (Zacek *et al.*, 1992). Although still available, it is not widely used by the poultry industry.

Importance of vaccination

MD vaccines were the first successful anti-cancer vaccines to be developed, and a number of other features make them unique in the field of vaccinology:

1. Most are sold as frozen cell-associated stocks requiring liquid nitrogen for transport and storage
2. MD vaccines should provide protection against very early challenge (within the first days of life), and therefore are administered at hatching or *in ovo* (see below)
3. A single administration should induce lifelong protection.

In spite of these difficulties, MD vaccines have been extremely effective at reducing MD losses. Losses due to condemnations for MD lesions in young broiler chickens in the USA have decreased from 1.5 per cent in 1970 to 0.0121 per cent in 1999 – a reduction of over 99 per cent (Witter, 2001). The use of MD vaccines is essential for the worldwide poultry industry; stopping MD vaccination would have huge economic consequences.

All licensed vaccines protect against mortality, clinical signs and gross MD lesions, but none of them provides protection against infection, replication and shedding of the challenge virus. Consequently, most poultry houses are infected with MDV. In areas of moderate contamination the incubation period for MD (7–10 weeks) is longer than the lifespan of broilers (6–7 weeks), and therefore they are not vaccinated; only birds with a longer lifespan (such as some free-range certified organic chicken, layers and breeders) are vaccinated. By contrast, in heavily infected and densely populated areas, where MDV strains of high virulence occur (e.g. the USA), broilers are vaccinated, usually with the HVT or with the bivalent HVT+MDV-2 vaccine. The *in ovo* route for broiler vaccination is mainly used in the USA (more than 95 per cent of broilers), but its use is increasing in Asia (Taiwan, Korea, Japan), Europe (mainly Spain and Italy) and South America (Argentina, Brazil). Most breeders and layers are vaccinated with the Rispens vaccine strain either alone or in combination with other vaccine serotypes (commonly HVT). In some countries, Rispens vaccine is also used for turkeys.

The importance of vaccination for the poultry industry is well illustrated by the situation in the USA (Witter, 1997). The introduction of the first HVT vaccine (1971) resulted in a rapid decline in the percentage of condemnations (Figure 13.1). However, several years later more pathogenic MDV strains (classified as very virulent, vvMDV) emerged in densely populated areas such as the Delmarva Peninsula in the USA. The new bivalent HVT+MDV-2 vaccine was licensed in 1983 and use of this bivalent vaccine brought condemnations down to acceptable levels, but again several years later a new pathotype (designated vv+MDV) appeared in the same areas. The Rispens vaccine was introduced in the mid-1990s and provided superior protection against the newly isolated vv+MDV pathotype. The condemnation trend line clearly shows not only the

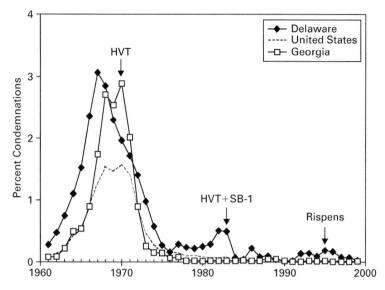

Figure 13.1   The reduction in the rates of condemnations of broiler chickens in the USA by the use of increasingly effective vaccines. (From Witter, R. L. (2001) *Current Topics in Microbiology and Immunology* (ed. K. Hirai), pp. 58–90. ©Springer-Verlag, Berlin, with permission.)

crucial role MD vaccines played in reducing MD losses, but also the risk of selecting highly virulent strains due to the pressure of vaccination. Although highly virulent MDV strains also have been identified in other countries, it is less clear if intensive vaccination has been responsible for the emergence of such strains.

Licensing, manufacturing and administering current MD vaccines

In order to become licensed, MD vaccines have to be tested in numerous studies based on the regulatory requirements of the country in which they are to be sold (e.g. 9CFR §113.130 in the USA and EU Pharmacopoeia for Europe). The master seed virus (MSV) is the seed-lot used to produce all the production batches, and is required to be free from defined extraneous agents. The MSV is tested for safety in chickens by administering at least 10 doses per bird and evaluating the outcome after 120 days by examining for gross MD lesions. In some countries, safety is also tested by examining virus recovered from five successive passages in chickens (not required for HVT by European Pharmacopoeia). The vaccine is tested for its immunogenicity in a challenge model specified by the regulatory requirements. It is notable that in the USA challenge requirements are more stringent for the MDV-1-derived vaccines. These must have higher efficacy against vvMDV than HVT vaccines. Each batch of vaccine needs to be tested for its purity, safety and potency (based on titre) before release. The titre of one dose is usually higher than the minimum required (at least 1000 plaque forming units (PFU) per dose in the USA).

The HVT, bivalent HVT+MDV-2 and CVI988 vaccines are the most widely-used MD vaccines today around the world. These vaccine strains are generally produced by infecting primary chicken embryo cells (CEC) obtained from specific-pathogen-free flocks and grown in roller bottles. At an optimal time after infection, the infected cells are harvested, concentrated, distributed in glass ampoules, slowly frozen in a specific medium containing a cryopreservative such as dimethyl sulphoxide (DMSO), and stored and transported in liquid nitrogen ($-196°C$). Each ampoule usually contains 1000 or 2000 doses of vaccine. Shortly before its administration, the vaccine must be thawed quickly and diluted in the provided diluent. The correct handling of MD vaccine is crucial for maintaining its efficacy. Only the HVT vaccine is also available as a freeze-dried product, and this has the advantage of being able to be stored at 4°C.

The vaccine is administered by either the subcutaneous route (0.2 ml/dose in the nape of the neck) or intramuscularly (leg) in 1-day-old chicks, usually at the hatchery. This vaccine is administered using a semi-automated device developed to expedite the process, so that one operator can deliver vaccine to between 2000–3000 chicks per hour (Witter, 2001). In the USA and some other countries broilers are vaccinated *in ovo*, approximately 3 days before hatching (see below). Vaccines for *in ovo* administration are identical to those used for administration at 1 day of age, except that they are suspended in a smaller volume of diluent and the volume administered to each egg is smaller.

The manufacturing of MD vaccine requires large numbers of specified pathogen-free (SPF) eggs, and is a laborious process involving production of CEC and handling of a large number of roller bottles. Cell lines such as the DF-1 cell line (Himly *et al.*, 1998), the OU2 cell line (Abujoub and Coussens, 1995; Abujoub *et al.*, 1999) and a quail cell line expressing MDV gE (Schumacher *et al.*, 2002) have been, or are currently being, tested for MD vaccine production to replace the need for SPF eggs. Attempts to produce MD vaccines in biofermenters have so far been unsuccessful.

In ovo vaccination

Within the last decade there has been a dramatic change in the method of delivery of MD vaccines in commercial broiler chickens. Previously, MD vaccines were administered at hatching by the subcutaneous route. Today, most major commercial hatcheries use the *in ovo* delivery system. With this system, live vaccine viruses are administered to embryonated eggs before hatching.

Injection *in ovo* is given at the time eggs are transferred from the incubator to the hatcher, usually around embryonation day (ED) 18. Automated, multiple-head injectors deliver a precise quantity of vaccine simultaneously to an entire tray of eggs (Ricks *et al.*, 1999). This simultaneous inoculation of large numbers of eggs saves on the labour costs associated with injecting individual chicks after hatching. The automatic injectors deposit the vaccine inoculum into the amniotic fluid of the majority of the eggs. There is no apparent adverse effect from *in ovo* injection on either the hatchability of the eggs or the long-term performance of the chickens. Chicks hatching from vaccinated eggs resist challenge with virulent MDV at hatching. The long-term protective efficacy of *in ovo* vaccination is comparable to that of parenterally administered vaccine after hatching. Further advances to refine and expand the *in ovo* delivery system for poultry are in progress (Sharma and Ricks, 2002).

The concept of *in ovo* vaccination was first examined by inoculating MDV-3 vaccine into SPF chicken eggs (Sharma and Burmester, 1982). Subsequent studies revealed that MDV-1 and -2 vaccines, and vaccines against a number of other pathogens, could be administered *in ovo* without compromising either the hatchability or well-being of the hatched chickens (Sharma, 1986; Wakenell and Sharma, 1986; Ahmad and Sharma, 1992; Noor *et al.*, 1995; Reddy *et al.*, 1996; Stone *et al.*, 1997; Karaca *et al.*, 1998). Recently, it was shown that a single injection of a multivalent *in ovo* vaccine (MIV) containing five live vaccine viruses was effective in immunizing chickens against multiple pathogens (Sharma *et al.*, 2002). The MIV comprised MDV-1, -2 and -3, a vaccine strain of serotype-1 infectious bursal disease virus (IBDV) and a recombinant fowlpox virus (FPV) containing HN and F gene inserts of Newcastle disease virus (NDV). Chickens given MIV at ED 18 developed specific antibodies against all the viral agents present in the mixture, and were protected against challenge with virulent MDV, IBDV, NDV and FPV. The performance of MIV-vaccinated commercial broilers was compared to that of hatchmates given only MDV-3 as an *in ovo* vaccine. Relative values for hatchability of the eggs, livability and weight gain of chickens and condemnation rates at processing were comparable between the MIV and the MDV-3 groups ($P > 0.05$). The successful use of MIV under field conditions indicates that the developing chicken embryo is able to make immune responses simultaneously to a number of infectious agents. This observation should expand the usefulness of the *in ovo* vaccination technology for poultry.

The mechanism by which *in ovo* vaccination induces protection against pathogen challenge requires elucidation, although it is very likely that the vaccine-induced protection in the embryo is mediated by the immune system (Sharma and Ahmad, 1995). Avian embryos, similar to mammalian foetuses, can develop a tolerance to infectious or non-infectious agents. The stage of embryonal development at the time of antigenic exposure determines the outcome of the response. Generally, if exposure occurs after the immune system has reached a critical stage of functional development, an immune response rather than tolerance is the likely outcome. The ability of MDV-3 to induce tolerance in chickens following *in ovo* vaccination was examined recently (Zhang and Sharma, 2003). Chickens exposed to MDV-3 vaccine at ED 14 or earlier showed 6–33 per cent incidence of tolerance to this virus. Tolerant chickens lacked detectable anti-MDV antibodies, developed persistent MDV-3 viraemia and had reduced resistance to virulent MDV. Exposure to MDV-3 after ED 14 did not induce detectable tolerance.

Preliminary attempts to understand the host–virus relationships in the embryo have revealed that introduced live viruses establish infection in embryonic tissues. However, the nature of infection can vary between viruses. For example, there was a marked difference in the latent infection induced in embryonic cells by MDV-1 and -3. MDV-3 replicated extensively in the embryo, and latently infected cells readily transmitted infectious virus when co-cultured with CEC *in vitro* (Sharma *et al.*, 1984). Embryonic cells latently infected with MDV-1, on the other hand, were unable to transfer infectious virus to co-cultured CEC (Sharma, 1987; St Hill and Sharma, 2000). However, if embryonic cells latently infected with MDV-1 were incubated *in vitro* for 48 hours or longer prior to co-culture with CEC, the virus was readily transferred to these cells (Y. Zhang, M. S. Parcells, J. T. Huyng, A. Copal, J. M. Sharma, personal communication). Thymus cells from chickens exposed to MDV-1 after hatching readily transferred this virus to CEC without prior *in vitro* incubation. This indicates that the nature of MDV latency in the embryonic cells is fundamentally different from that established in cells of hatched chickens. The mechanism(s) of MDV-induced latency in chicken cells clearly needs further study.

In addition to inducing latent infection in embryonic cells, *in ovo* exposure to MDV-1 can also result in expression of viral antigens. Chicken eggs were inoculated at ED 17 with a recombinant MDV containing a green fluorescent protein (GFP) gene insert. This recombinant, designated MDV-RB1BUS6smGFP, was constructed using an smGFP expression vector cassette inserted at a unique *Avr* II site within the US6 coding region of MDV (Parcells *et al.*, 1995, 2001). Embryonic tissues were examined at intervals for GFP expression. By 4 days post-inoculation, fluorescence was detected in a number of tissues, most abundantly in the thymus (Zhang *et al.*, 2003). It is likely that viral proteins were also expressed along with GFP. The more compelling evidence of viral expression in the embryo comes from challenge studies. SPF chickens were immunized by *in ovo* vaccination with the CVI988 strain of serotype 1 MDV. Chicks hatching from vaccinated eggs were resistant to challenge with virulent MDV at hatching (Zhang and Sharma, 2001), indicating that the viral proteins were adequately expressed in the embryo to initiate a protective response.

Vaccination problems and solutions

Vaccine failures have been reported in the field and can be caused by many different factors (see also Chapter 5). The maintenance of the cold chain during the storage and transport of cell-associated herpesvirus is an absolute necessity for good efficacy, and may be difficult to achieve in some developing countries. The preparation (thawing and dilution) of the vaccine is also critical. The diluent provided by the manufacturer needs to be used without any other additives (antibiotics or other vaccines), and the recommended dilution must be used. The vaccine needs to be delivered shortly after dilution, using a clean and adequate system of administration. If all of these handling steps are not properly carried out, then the dose of the vaccine administered to the chick or embryo may be much lower than a minimum protective dose, and this may result in vaccine failure. Although different doses of vaccines induce similar levels of protection in controlled studies, under field conditions higher doses of vaccine establish earlier infection and more effectively overcome maternal antibodies (Witter, 2001). In some countries MD vaccines are deliberately over-diluted, and this practice may cause vaccine failure.

Vaccine failures can be due to challenge from highly virulent MDV strains (such as vv+MDV), especially in areas of high chicken density. In vv+MDV-infected areas, it is very important to improve management practices and biosecurity measures to reduce the level of environmental contamination and to delay and minimize the vv+MDV exposure dose. If this is not successful, use of a bivalent (Rispens+HVT) or trivalent (Rispens+MDV-2+HVT) vaccine can help. Additionally, two doses of vaccine delivered on the same day, or 1–3 weeks apart, have been shown to improve the situation in the field, even if an advantage cannot be demonstrated in experimental studies (Witter, 2001). The use of multivalent vaccines can compensate for the lower protection induced by HVT and the delayed protection induced by CVI988 (Geerligs *et al.*, 1999). However, these highly effective vaccines should not be used routinely in broilers in order to reduce the risk of emergence of MDV strains of higher virulence.

As with genetic susceptibility to MD (see Chapter 9), the genetic background of the chicken recipient has an influence on vaccine efficacy. Bacon and Witter (1993, 1994) showed that the influence of the MHC haplotype on vaccine efficacy is dependent on the serotype of the vaccine concerned. Wakenell *et al.* (1996) reported differences between $B^{15/19}$ chickens and commercial SPF birds in protection induced by HVT vaccine or a recombinant HVT expressing MDV genes. Detailed understanding of the basis for such differences could pave way to the development of genotype-specific vaccines in the future (Witter, 2001).

Vaccine efficacy can also be influenced by the health status of chickens, and their rearing environment. As with any other vaccine, the immune system of the recipient needs to be fully functional for a vaccine to be effective. Stress or a concomitant infection can have a negative effect on vaccine uptake. In particular, infection with agents that interfere with the immune system (such as IBDV, reticuloendotheliosis virus (REV), reovirus, or chicken infectious anaemia virus) have been shown to inhibit MD vaccine-induced immunity (see Chapter 11). Increasing biosecurity measures, such as adopting an 'all-in all-out' system of management with cleaning and disinfection between flocks (rather than placing chicks on built-up litter), and avoiding contact between chickens of different ages, will reduce the pressure of infection and losses from MD.

Maternal antibodies against the different MDV serotypes reduced the level of viraemia and protection induced by vaccination. This effect is particularly pronounced with cell-free HVT vaccines (Yoshida *et al.*, 1975; Prasad, 1978; Witter and Burmester, 1979) but it has been observed as well with cell-associated ones (Witter and Lee, 1984). In order to decrease the inhibiting effect of maternal antibodies, it is recommended to vaccinate breeders with vaccine (e.g. MDV-1) different from that (e.g. HVT) used in the progeny (King *et al.*, 1981).

Attenuation of MDV and its effect on protection

Pathogenic MDV-1 can be attenuated by serial passage in cell culture. The number of passages required for attenuation is variable, and depends on the strain of MDV and the criteria used to evaluate attenuation (Witter, 2001). The first MD vaccine developed was based on a virulent MDV-1 isolate (HPRS-16) attenuated by 33 passages on chicken kidney cells. However, Witter (1982, 2002) needed to passage strains of vvMDV (Md11) or vv+MDV (648A and 584A) 70–100 times using CEC in order to achieve complete attenuation. Such passages may have included plaque purification. This was done to generate clone C of the CVI988 strain, which, in contrast to the parental strain (von Bülow, 1977), was safe in highly susceptible chicken such as SPF Rhode Island Reds (Pol *et al.*, 1986). The resulting attenuated viruses usually induce large plaques and grow to higher titres than the parental virus in cell culture (see Chapter 12). However, they may lose the ability to replicate well in chickens, which can be the result of loss of lymphotropism (Schat *et al.*, 1985). This occurred with CVI988 clone C and Md11/75C strains, which were poorly protective. In order to enhance replication *in vivo*, these isolates were back-passaged several times in chickens to generate the CVI988C/R6 (de Boer *et al.*, 1986, 1987) and Md11/75C/R2 strains (Witter, 1987, 1991), respectively. The latter was still pathogenic in SPF chickens and was further attenuated by *in vitro* passages to generate the Md11/75C/R2/23 vaccine strain, which was licensed in the USA (Witter *et al.*, 1995). Thus, safe MDV-1-derived vaccine strains can be generated by cell culture passage, and excessive attenuation can be partly reversed by back-passage in chickens.

In maternal antibody-positive (Ab$^+$) chickens, protection induced by partially attenuated strains of vv+MDV (e.g. passage 80 of 648A strain) is greater than that induced by fully attenuated strains (e.g. passage 100 of 648A), and this correlates with levels of viraemia (Witter, 2002; Gimeno *et al.*, 2004). Although partially attenuated strains of vv+MDV caused gross lymphomas or enlarged nerves in some susceptible, maternal antibody-negative (Ab$^-$) chickens, they were completely safe in Ab$^+$ chickens. Relationships therefore

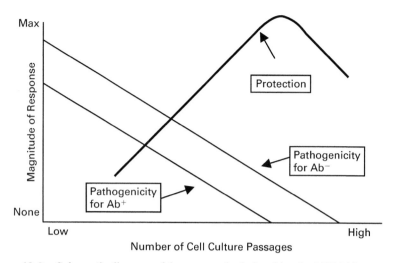

Figure 13.2 Schematic diagram of the proposed relationships for MDV-1 between the number of serial passages in cell culture, the ability to induce protective immunity, and the pathogenicity for 1-day-old antibody negative (Ab^-) or positive (Ab^+) chickens. Responses are the relative frequency of gross MD lesions after vaccination (pathogenicity) and the proportional inhibition of gross lesions after challenge of vaccinated chickens (protection). (From Witter, R. L. (2002) Induction of strong protection by vaccination with partially attenuated serotype 1 Marek's disease viruses. *Avian Dis.*, **46,** 925–937, with permission.)

exist between passage level, virulence and the protective ability of attenuated MDV-1; these are illustrated in Figure 13.2 (Witter, 2002). Optimal protection is induced at a passage level that is safe for Ab^+ chickens, but not for Ab^- chickens. The challenge for developing new attenuated vaccine strains is to find the optimal balance between vaccine efficacy and safety. If more efficacious vaccines are needed, they will probably be less safe (in Ab^- chickens). 'Hot' vaccine strains have been developed for other poultry pathogens such as IBDV. However, the use of 'hot' vaccine should be restricted to those situations where good management practices and adequate biosecurity measures have failed to control an outbreak.

The genetic bases of attenuation are not understood. The fact that attenuation requires many passages suggests that attenuation of MDV, as for other herpesviruses, is the result of multiple mutations at different loci in the genome (Witter, 2001). Viral populations with different mutations will be produced at each passage in cell culture and, progressively, mutants with growth advantages will emerge as the major virus population. These mutants will have mutations responsible for the attenuated phenotype, as well as other co-segregating mutations. Moreover, the genetic changes responsible for attenuation in different MDV strains are likely to be different. It is therefore very difficult to identify those genetic changes responsible for causing attenuation in a particular vaccine strain.

Different attenuated strains obtained by repeated cell culture passage were found to have lost the expression of the MDV A antigen (the homologue of herpes simplex glycoprotein C) (Churchill *et al.*, 1969; Witter and Offenbecker, 1979; Wilson *et al.*, 1994). However, the CVI988 strain retained its A antigen (Rispens *et al.*, 1972a), indicating that the loss of A antigen is not essential for attenuation. A 132 base pair (bp) repeat located in the *Bam*HI-H fragment was identified as 6–35 copies in attenuated strains compared to 2–3 copies in virulent strains (Maotani *et al.*, 1986). Several authors later showed that the increase and decrease in number of the 132 bp repeats occurred during *in vitro* and *in vivo* passages, respectively, and were not responsible for attenuation (Ross *et al.*, 1993a; Hooft van Iddekinge *et al.*, 1999; Silva *et al.*, 2004). Other differences in DNA sequence or in transcription profiles between virulent and attenuated strains have been detected by different techniques (Schat *et al.*, 1989; Wilson and Coussens, 1991; Ross *et al.*, 1993a; Liu *et al.*, 1996; Niikura *et al.*, 1996; Jarosinski *et al.*, 2003). It is not known if these differences are responsible for attenuation.

All attempts to produce a better vaccine than the CVI988 strain isolated some 30 years ago have so far failed – except, perhaps, for a recent Australian MDV vaccine (Karpathy *et al.*, 2002, 2003). This suggests that the CVI988 strain has some unique protective ability that cannot easily be obtained by attenuation of virulent isolates. Identifying the genetic basis for making this strain more protective than the other MDV-1 candidates should help to develop more efficacious vaccines in future. The Rispens strain contains a unique mutation in the pp38 gene, which prevents this strain from reacting with the H19 monoclonal antibody. An H19-positive CVI988 mutant, generated by recombination, induced delayed and lower antibody titres against MDV-1 compared with the parental CVI988 strain. This suggests that a unique mutation at the H19 epitope of pp38 in the CVI988 strain could be one of the factors responsible for the superior protection of this vaccine strain (Cui *et al.*, 1999). Additional data (e.g. with a rescue mutant) are needed to confirm that the phenotype of the mutant is due to a unique amino acid mutation in pp38. As mentioned above, protection levels induced by different vaccine candidates often correlate with those of vaccine-induced viraemia. However, a SORF2-GFP attenuated RB1B mutant induced a higher level of viraemia and a lower level of protection compared to the CVI988 strain (M. Bublot, N. Pritchard, K. Karaca, A. Robles, J. Cruz and M. S. Parcells, personal communication). This observation suggests that factors other than viraemia are responsible for the high performance of CVI988 vaccine. The CVI988 strain has been recently cloned as bacterial artificial chromosome (BAC) clones (Petherbridge *et al.*, 2003), and its sequence is likely to be published soon. The BAC technology should facilitate the difficult task of finding sequence differences responsible for the biological properties of CVI988.

Another example of the attenuation and protection phenomenon is the RM1 clone. This clone was derived from the JM strain of vMDV-1 through retrovirus insertional mutagenesis; it contains sequences from REV inserted at the junction of the internal repeat and unique short (U_S) regions of the MDV genome. RM1 was attenuated for oncogenicity but caused early cytolytic infection, illustrated by marked thymic and bursal atrophy. Levels of protection against vv+MDV challenge induced by RM1 exceeded those of CVI988 (Witter *et al.*, 1997). Further studies are needed to determine if this protection was due either to the RM1-induced destruction of primary target cells for MDV or to other properties of the RM1 clone.

Recombinant vaccines

A recombinant virus is one that has been genetically modified by recombination. It usually contains a foreign sequence inserted into a particular locus within the genome, and this insertion may result in the deletion of a small segment of the genome. If the locus is within a genomic region involved in virulence, the insertion of the foreign sequence can result in an attenuated phenotype. The inserted foreign sequence is often a complete expression cassette containing the foreign gene and/or a marker gene (such as GFP). The foreign gene will be expressed in cells infected with the recombinant virus, and its product will be recognized by the host's immune system or, if it is an immunomodulatory gene, can modulate the immune system of the host. Classically, recombinant viruses are generated by co-transfection of the parental genome with a donor plasmid containing the foreign sequence flanked by sequences identical to those flanking the locus of insertion. However, this low efficiency and time-consuming strategy is being replaced by new technologies involving either overlapping cosmids or BAC that have been successfully adapted for MDV (Reddy *et al.*, 2002 and Schumacher *et al.*, 2000, respectively).

Different types of recombinant viruses have been proposed as MD vaccine candidates (Hirai and Sakaguchi, 2001). Nazerian *et al.* (1992, 1996) generated and tested FPV recombinants expressing different MDV genes, including: gB, gC, gD, UL47 and UL48. Only the gB recombinant provided an excellent level of protection in SPF chicks, and gB from MDV-1 was superior to that from other serotypes. However, in Ab+ broilers protection was very poor, indicating that gB is a likely target for maternal antibodies. Similar findings with gB were reported by Heine *et al.* (1997) and Liu *et al.* (1999). Interestingly, a synergism between fowlpox/gB and HVT (Nazerian *et al.*, 1996; Liu *et al.*, 1999) was found. Additional fowlpox recombinants expressing gE, gI, gH and UL32 were recently evaluated in Ab+ chickens (Lee *et al.*, 2003). Only the gI recombinant induced a level of protection (approx. 40 per cent) comparable to that induced by the gB recombinant or HVT. A synergism (approx. 70 per cent protection) was observed with fowlpox recombinants expressing multiple genes including gB, gI and gE with or without UL32. A synergism (94 per cent protection) was also observed when the gB/gI/gE/UL32 recombinant was associated with HVT. Both of these viruses can be freeze-dried, and their combination is the best 'cell-free' MD vaccine so far described. A double fowlpox recombinant expressing both MDV gB and chicken interferon-γ was recently shown to induce better protection than the single MDV gB construct (Liu *et al.*, 2000). An FPV recombinant expressing chicken myelomonocytic growth factor (cMGF) prolonged survival time and reduced viraemia and tumour incidence of chickens challenged with MDV. Furthermore, it significantly improved the protection induced by HVT (Djeraba *et al.*, 2002). Expression of immunomodulatory genes in a FPV vector may therefore increase the efficacy of classical and recombinant MD vaccines.

In order to improve MD protection induced by HVT, Ross *et al.* (1993b) generated an HVT recombinant expressing the gB from MDV-1. This gB gene was inserted into the TK gene of HVT, and the insertion resulted in a lower ability to replicate *in vivo* compared with the parental HVT. Nevertheless, this recombinant induced a higher level of protection at high doses (5000 PFU) than a TK negative HVT, indicating that the MDV-1 gB gene improved protection (Ross *et al.*, 1996). Currently, two HVT recombinants expressing MDV genes (one expressing gB, gC and gD and the other expressing gB and gC as well as NDV genes encoding the HN and F proteins) have been licensed in the USA (Reddy *et al.*, 1996). However, published data did not show a clear positive effect of MDV gene expression on HVT-induced protection, and those recombinants were not launched for use in the broiler industry. Improving the efficacy of the HVT vaccine was recently attempted by generating HVT recombinants expressing chicken interleukin-2 (Tarpey *et al.*, 2002) and CD30 (Burgess *et al.*, 2002), a tumour associated antigen over-expressed in MD lymphomas (see Chapter 8). Both of these recombinants induced levels of MD protection similar to that induced by the parental HVT.

Another strategy for producing improved MD vaccines using recombinant technology is to generate a genetically attenuated mutant from a virulent MDV isolate by inactivating genes whose products are critical for early cytolytic infection and tumour formation, or which interact negatively with the immune system but do not significantly affect viral replication. In order to identify such genes, targeted deletion is performed and the phenotype of the resulting deletion mutant evaluated. Studies of different vvMDV recombinants have allowed identification of genes involved in virulence (e.g. Parcells *et al.*, 1995; Morgan *et al.*, 1996; Anderson *et al.*, 1998; Parcells *et al.*, 2001; Reddy *et al.*, 2002, 2003). The vvMDV RB1B strain has been cloned into a BAC vector (Petherbridge and Nair, 2002). BAC technology should considerably speed up determination of the roles that the different MDV genes play in pathogenesis. Genetically attenuated vaccines are generally more stable than viruses that have been attenuated by cell passage. However, multiple mutations, deletions or insertions need to be introduced into different parts of the genome (unique small (U$_S$) and unique long (U_L) regions) in order to avoid generation of more virulent viruses by recombination with field isolates, as described for other herpesviruses (e.g. Javier *et al.*, 1986).

The latest generation of recombinant MD vaccine candidates is based on vaccination with DNA containing the whole genome of MD cloned into a BAC vector (Tischer *et al.*, 2002; Petherbridge *et al.*, 2003). The BAC clone DNA is produced in *Escherichia coli*. As with other herpesviruses, the genomic DNA of MDV is infectious, and recovery of virus can easily be obtained by transfecting sensitive cells with the BAC clone DNA. Vaccination of chickens with BAC clone DNA induced partial protection against challenge with vvMDV. The reconstitution of replication-competent MDV vaccine was critical for protection, because the non-replicative gE-negative mutant, or its BAC DNA, did not induce protection (Tischer *et al.*, 2002). These new types of vaccines have the important advantages of being molecularly defined, stable at 4°C, and able to be grown in *E. coli*. However, one major drawback compared to existing vaccines is likely to be the delayed onset of protection, due to the time required to generate sufficient infectious virus from the DNA. Enhancing the uptake of DNA and reconstituting replicative virus will be necessary for these vaccines to be successful. A further improvement would be to develop a bacteria vector capable of replicating the BAC clone and transmitting it to other cells in the host, as recently described for murine cytomegalovirus (Cicin-Sain *et al.*, 2003).

HVT has been used as a vector to express protective genes of other avian pathogens such as NDV (Morgan *et al.*, 1992, 1993; Sondermeijer *et al.*, 1993; Heckert *et al.*, 1996; Reddy *et al.*, 1996), IBDV (Darteil *et al.*, 1995; Bublot *et al.*, 1996, 1999; Tsukamoto *et al.*, 2002), infectious laryngotracheitis virus (Saif *et al.*, 1993), *Eimeria* (Cronenberg *et al.*, 1999), and avian leukosis virus type J (A. M. Fadly *et al.*, East Lansing, personal communication). MDV has also been developed as a vector for pathogens such as IBDV (Tsukamoto *et al.*, 1999) and NDV (Sakaguchi *et al.*, 1998; Sonoda *et al.*, 2000). Important advantages of these vectors include the following:

1. They can be administered by the *in ovo* route
2. They provide protection against MD as well as the other diseases, depending on the protective genes introduced
3. They are only weakly susceptible to interference from maternal antibody
4. They induce long-lasting protection
5. Vaccinated birds can be distinguished from infected birds.

One of the major disadvantages is that they provide only poor mucosal immunity. This is illustrated by the HVT/NDV recombinant, which protects chickens against NDV-induced mortality but not against NDV replication in the trachea (Morgan *et al.*, 1992; Reddy *et al.*, 1996). Nevertheless, HVT/IBDV recombinant vaccines are in a late stage of development in Europe and the USA. The major advantages of recombinant IBDV vaccines over the conventional vaccines are the lack of bursal damage and their efficacy in maternal antibody-positive chickens. Recombinant vaccines have also been shown to be effective in prime–boost immunization strategies (Tsukamoto *et al.*, 2000; Kross-Landsman *et al.*, 2003).

Recombinant vaccines need to show clear advantages over existing vaccines in order to be widely accepted by the poultry industry. Recombinant technology used to design better MD vaccines has been disappointing so far, but new tools are becoming available that should speed up the generation of numerous vaccine candidates. The limiting factor is likely to be evaluation of the safety and efficacy of candidate vaccines in the chicken. The use of HVT or MDV as a vector, to carry the protective genes from other avian pathogens such as IBDV, looks very promising and, under field conditions, could provide distinct advantages over existing vaccines.

Summary

MD vaccines, first introduced in the 1970s, have brought enormous benefits to poultry production and without them the modern poultry industry would not have survived the scourge of MD and been able to become so productive. Almost all layers and breeders are vaccinated against MD and in some countries where MD is a problem, broilers are also vaccinated. Of the MD vaccines licensed during the last 30 years or so, the CVI988 (Rispens) vaccine has become the 'gold standard' providing superior protection against highly virulent challenge strains of MDV. Nevertheless, MD outbreaks continue to be reported in several countries around the world and improved vaccination strategies have been developed – using of multivalent vaccines, *in ovo* vaccination or revaccination. However, MD vaccines are produced in CEC from SPF chickens, which is costly and leaves open the risk of other agents being introduced via MD vaccines. None of the current vaccines engender sterilizing immunity and vaccinated chickens can still be infected with highly virulent MDV that can replicate, be shed and infect others in the flock. This situation has almost certainly contributed to the problems of field strains of MDV increasing virulence, as illustrated by the current situation in the USA (Witter, 1997). New research using recombinant technology now presents the opportunity for developing new vaccines which are sustainable and may have superior protective properties (see Chapter 14).

References

Abujoub, A. A. and Coussens, P. M. (1995). *Virology*, **214**, 541–549.
Abujoub, A. A., Williams, D. L. and Reilly, J. D. (1999). *Acta Virol.*, **43**, 186–191.
Ahmad, J. and Sharma, J. M. (1992). *Am. J. Vet. Res.*, **53**, 1999–2004.
Anderson, A. S., Parcells, M. S. and Morgan, R. W. (1998). *J. Virol.*, **72**, 2548–2553.

Bacon, L. D. and Witter, R. L. (1993). *Avian Dis.*, **37**, 53–59.
Bacon, L. D. and Witter, R. L. (1994). *Avian Dis.*, **38**, 65–71.
Bacon, L. D., Witter, R. L. and Fadly, A. M. (1989). *J. Virol.*, **63**, 504–512.
Bublot, M., Laplace, E., Bouquet, J.-F. *et al.* (1996). In *Current Research on Marek's Disease* (eds R. F. Silva, H. H. Cheng, P. M. Coussens *et al.*), pp. 402–407. American Association of Avian Pathologists, Kennett Square, Pennsylvania.
Bublot, M., Laplace, E. and Audonnet, J.-C. (1999). *Acta Virol.*, **43**, 181–185.
Burgess, S. C., Baaten, B. J. G., Baxendale, W. *et al.* (2002). *Workshop on Molecular Pathogenesis of Marek's Disease and Avian Immunology, Limassol, Cyprus*, p. 40, Kimron Institute, Beit Dagan, Israel.
Calnek, B. W., Schat, K. A., Peckham, M. C. and Fabricant, J. (1983). *Avian Dis.*, **27**, 844–849.
Churchill, A. E. and Biggs, P. M. (1967). *Nature*, **215**, 528–530.
Churchill, A. E., Payne, L. N. and Chubb, R. C. (1969). *Nature*, **221**, 744–747.
Cicin-Sain, L., Brune, W., Bubic, I. *et al.* (2003). *J. Virol.*, **77**, 8249–8255.
Cronenberg, A. M., van Geffen, C. E. H., Dorrestein, J. *et al.* (1999). *Acta Virol.*, **43**, 192–197.
Cui, Z., Qin, A., Lee, L. F. *et al.* (1999). *Acta Virol.*, **43**, 169–173.
Darteil, R., Bublot, M., Laplace, E. *et al.* (1995). *Virology*, **211**, 481–490.
de Boer, G. F., Groenendal, J. E., Boerrigter, H. M. *et al.* (1986). *Avian Dis.*, **30**, 276–283.
de Boer, G. F., Pol, J. M. A. and Oei, H. I. (1987). *Vet. Q.*, **9** (Suppl. 1)**,** 16–28.
Djeraba, A., Musset, E., Lowenthal, J. W. *et al.* (2002). *J. Virol.*, **76**, 1062–1070.
Geerligs, H. J., Weststrate, M. W., Pertile, T. L. *et al.* (1999). *Acta Virol.*, **43**, 198–200.
Gimeno, I. M., Witter, R. L., Hunt, H. D. *et al.* (2004). *Avian Pathol.*, **33**, 59–68.
Heckert, R. A., Riva, J., Cook, S. *et al.* (1996). *Avian Dis.*, **40**, 770–777.
Heine, H. G., Foord, A. J., Young, P. L. *et al.* (1997). *Virus Res.*, **50**, 23–33.
Himly, M., Foster, D. N., Bottoli, I. *et al.* (1998). *Virology*, **248**, 295–304.
Hirai, K. and Sakaguchi, M. (2001). *Curr. Topics Microbiol. Immunol.*, **255**, 261–287.
Hooft van Iddekinge, B. J. L., Stenzler, L., Schat, K. A. *et al.* (1999). *Avian Dis.*, **43**, 182–188.
Jarosinski, K. W., O'Connell, P. H. and Schat, K. A. (2003). *Virus Genes*, **26**, 255–269.
Javier, R. T., Sedarati, F. and Stevens, J. G. (1986). *Science*, **234**, 746–748.
Karaca, K., Sharma, J. M., Winslow, B. J. *et al.* (1998). *Vaccine*, **16**, 1–7.
Karpathy, R. C., Firth, G. A. and Tannock, G. A. (2002). *Aust. Vet. J.*, **80**, 61–66.
Karpathy, R. C., Firth, G. A. and Tannock, G. A. (2003). *Aust. Vet. J.*, **81**, 222–225.
King, D., Page, D., Schat, K. A. and Calnek, B. W. (1981). *Avian Dis.*, **25**, 74–81.
Kross-Landsman, I., Koolen, M. J. M., Sondermeijer, P. J. A. and Mebatsion, T. (2003). In *XIII Congress of the World Veterinary Poultry Association, Denver*, pp. 107–108, AAAP, Kennett Square, Pennsylvania.
Lee, L. E., Witter, R. L., Reddy, S. M. *et al.* (2003). *Avian Dis.*, **47**, 549–558.
Liu, H.-C., Silva, R. F. and Cheng, H. (1996). In *Current Research on Marek's Disease* (eds R. F. Silva, H. H. Cheng, P. M. Coussens *et al.*), pp. 148–153. American Association of Avian Pathologists, Kennett Square, Pennsylvania.
Liu, X., Peng, D., Wu, X. *et al.* (1999). *Acta Virol.*, **43**, 201–205.
Liu, X., Peng, D., Chen, J. *et al.* (2000). *6th International Symposium on Marek's Disease, August 20–23, Montréal*, MDS12.5, AAAP, Kennett Square, Pennsylvania.
Maas, H. J. L., Borm, F. and van de Kieft, G. (1982). *World's Poultry Sci. J.*, **38**, 163–175.
Maotani, K. A., Kanamori, A., Ikuta, K. *et al.* (1986). *J. Virol.*, **58**, 657–659.
Morgan, R. W., Gelb, J. Jr, Schreurs, C. S. *et al.* (1992). *Avian Dis.*, **36**, 858–870.

Morgan, R. W., Gelb, J. Jr, Pope, C. R. and Sondermeijer, P. J. (1993). *Avian Dis.*, **37**, 1032–1040.

Morgan, R., Anderson, A., Kent, J. and Parcells, M. (1996). In *Current Research on Marek's Disease* (eds R. F. Silva, H. H. Cheng, P. M. Coussens *et al.*), pp. 207–212. American Association of Avian Pathologists, Kennett Square, Pennsylvania.

Nazerian, K., Solomon, J. J., Wutter, R. L. and Burmester, B. R. (1968). *Proc. Soc. Exp. Biol. Med.*, **127**, 177–182.

Nazerian, K., Lee, L. F., Yanagida, N. and Ogawa, R. (1992). *J. Virol.*, **66**, 1409–1413.

Nazerian, K., Witter, R. L., Lee, L. F. and Yanagida, N. (1996). *Avian Dis.*, **40**, 368–376.

Niikura, M., Cheng, H. and Silva, R. F. (1996). In *Current Research on Marek's Disease* (eds R. F. Silva, H. H. Cheng, P. M. Coussens *et al.*), pp. 154–159. American Association of Avian Pathologists, Kennett Square, Pennsylvania.

Noor, S. M., Husband, A. J. and Widders, P. R. (1995). *Br. Poultry Sci.*, **36**, 563–573.

Okazaki, W., Purchase, H. G. and Burmester, B. R. (1970). *Avian Dis.*, **14**, 413–429.

Parcells, M. S., Anderson, A. S. and Morgan, R. W. (1995). *J. Virol.*, **69**, 7888–7898.

Parcells, M. S., Lin, S. F., Dienglewicz, R. L. *et al.* (2001). *J. Virol.*, **75**, 5159–5173.

Petherbridge, L. and Nair, V. (2002). *Workshop on Molecular Pathogenesis of Marek's Disease and Avian Immunology, Limassol, Cyprus*, pp. 48, Kimron Institute, Beit Dagan, Israel.

Petherbridge, L., Howes, K., Baigent S. J. *et al. J. Virol.*, **77**, 8712–8718.

Pol, J. M. A., Kok, G. L., Oei, H. L. and de Boer, G. F. (1986). *Avian Dis.*, **30**, 271–275.

Prasad, L. B. M. (1978). *Br. Vet. J.*, **134**, 315–321.

Reddy, S. M. and Lupiani, B. (2003). In *XIII Congress of the World Veterinary Poultry Association, Denver*, pp. 103, AAAP, Kennett Square, Pennsylvania.

Reddy, S. K., Sharma, J. M., Ahmad, J. *et al.* (1996). *Vaccine*, **14**, 469–477.

Reddy, S. M., Lupiani, B., Gimeno, I. M. *et al.* (2002). *Proc. Natl Acad. Sci. USA*, **99**, 7054–7059.

Ricks, C. A., Avakian, A., Bryan, T. *et al.* (1999). *Adv. Vet. Med.*, **41**, 495–515.

Rispens, B. H., van Vloten, J. and Maas, H. J. L. (1969). *Br. Vet. J.*, **125**, 445–453.

Rispens, B. H., van Vloten, H., Mastenbroek, N. *et al.* (1972a). *Avian Dis.*, **16**, 108–125.

Rispens, B. H., van Vloten, H., Mastenbroek, N. *et al.* (1972b). *Avian Dis.*, **16**, 126–138.

Ross, N., Binns, M. M., Sanderson, M. and Schat, K. A. (1993a). *Virus Genes*, **7**, 33–51.

Ross, N., Binns, M. M., Tyers, P. *et al.* (1993b). *J. Gen. Virol.*, **74**, 371–377.

Ross, N., O'Sullivan, G. and Coudert, F. (1996). *Vaccine*, **14**, 187–189.

Saif, Y. M., Rosenberger, J. K., Cloud, S. S. *et al.* (1993). *Proceedings of the 130th Annual Meeting of the American Veterinary Medical Association, Minneapolis, USA*, p. 154, AVMA, Schaumburg, Illinois.

Sakaguchi, M., Nakamura, H., Sonoda, K. *et al.* (1998). *Vaccine*, **16**, 472–479.

Schat, K. A. and Calnek, B. W. (1978). *J. Natl Cancer Inst.*, **60**, 1075–1082.

Schat, K. A., Calnek, B. W., Fabricant, J. and Grahm, D. L. (1985). *Avian Pathol.*, **14**, 127–146.

Schat, K. A., Buckmaster, A. and Ross, L. J. (1989). *Intl J. Cancer*, **44**, 101–109.

Schumacher, D., Tischer, B. K., Fuchs, W. and Osterrieder, N. (2000). *J. Virol.*, **74**, 11088–11098.

Schumacher, D., Tischer, B. K., Teifke, J. P. *et al.* (2002). *J. Gen. Virol.*, **83**, 1987–1992.

Sharma, J. M. (1985). *Avian Dis.*, **29**, 1155–1169.

Sharma, J. M. (1986). *Avian Dis.*, **30**, 776–781.

Sharma, J. M. (1987). *Avian Dis.*, **31**, 570–576.

Sharma, J. M. and Ahmad, J. (1995). In *Advances in Avian Immunology Research* (eds T. F. Davison, N. Bumstead and P. Kaiser), pp. 273–277. Carfax Publishing Co., Oxford.

Sharma, J. M. and Burmester, B. R. (1982). *Avian Dis.*, **26**, 134–149.

Sharma, J. M. and Ricks, C. (2002). *Proceedings of the American Association of Avian Pathologists Symposium, Nashville, Tennessee*, AAAP, Kennett Square, Pennsylvania.

Sharma, J. M., Lee, L. F. and Wakenell, P. S. (1984). *Am. J. Vet. Res.*, **45**, 1619–1623.

Sharma, J. M., Zhang, Y., Jensen, D. *et al.* (2002). *Avian Dis.*, **46**, 613–622.

Silva, R. F., Reddy, S. M. and Lupiani, B. (2004). *J. Virol.*, **78**, 733–740.

Sondermeijer, P. J., Claessens, J. A., Jenniskens, P. E. *et al.* (1993). *Vaccine*, **11**, 349–358.

Sonoda, K., Sakaguchi, M., Okamura, H. *et al.* (2000). *J. Virol.*, **74**, 3217–3226.

St Hill, C. A. and Sharma, J. M. (2000). *Avian Dis.*, **44**, 842–852.

Stone, H., Mitchell, B. and Brugh, M. (1997). *Avian Dis.*, **41**, 856–863.

Tarpey, I., Davis, P. J., Sondermeijer, P. *et al.* (2002). *Workshop on Molecular Pathogenesis of Marek's Disease and Avian Immunology, Limassol, Cyprus*, pp. 34, Kimron Institute, Beit Dagan, Israel.

Tischer, B. K., Schumacher D., Beer, M. *et al.* (2002). *J. Gen. Virol.*, **83**, 2367–2376.

Tsukamoto, K., Kojima, C., Komori, Y. *et al.* (1999). *Virology*, **257**, 352–362.

Tsukamoto, K., Sato, T., Saito, S. *et al.* (2000). *Virology*, **269**, 257–267.

Tsukamoto, K., Saito, S., Saeki, S. *et al.* (2002). *J. Virol.*, **76**, 5637–5645.

Vielitz, E. and Landgraf, H. (1971). *Dtsch Tierarztl. Wochenschr.*, **78**, 617–623.

von Bülow, V. (1977). *Avian Pathol.*, **6**, 395–403.

Wakenell, P. S. and Sharma, J. M. (1986). *Am. J. Vet. Res.*, **47**, 933–938.

Wakenell, P., Miller, M., Schwartz, R. and Chase W. (1996). In *Current Research on Marek's Disease* (eds R. F. Silva, H. H. Cheng, P. M. Coussens *et al.*), pp. 52–56. American Association of Avian Pathologists, Kennett Square, Pennsylvania.

Wilson, M. R. and Coussens, P. M. (1991). *Virology*, **185**, 673–680.

Wilson, M. R., Southwick, R. A., Pulaski, J. T. *et al.* (1994). *Virology*, **199**, 393–402.

Witter, R. L. (1982). *Avian Pathol.*, **11**, 49–62.

Witter, R. L. (1987). *Avian Dis.*, **31**, 752–765.

Witter, R. L. (1991). *Avian Dis.*, **35**, 877–891.

Witter, R. L. (1995). *Avian Pathol.*, **24**, 665–678.

Witter, R. L. (1997). *Avian Dis.*, **41**, 149–163.

Witter, R. L. (2001). In *Current Topics in Microbiology and Immunology* (ed K. Hirai), pp. 58–90. Springer-Verlag, Berlin.

Witter, R. L. (2002). *Avian Dis.*, **46**, 925–937.

Witter, R. L. and Burmester, B. R. (1979). *Avian Pathol.*, **8**, 145–156.

Witter, R. L. and Lee, L. F. (1984). *Avian Pathol.*, **13**, 75–92.

Witter, R. L. and Offenbecker, L. (1979). *J. Natl Cancer Inst.*, **62**, 143–151.

Witter, R. L., Nazerian, K., Purchase, H. G. and Burgoyne G. H. (1970). *Am. J. Vet. Res.*, **31**, 525–538.

Witter, R. L., Silva, R. F. and Lee, L. F. (1987). *Avian Dis.*, **31**, 829–840.

Witter, R. L., Lee, L. F. and Fadly, A. M. (1995). *Avian Dis.*, **39**, 269–284.

Witter, R. L., Li, D., Jones, D. *et al.* (1997). *Avian Dis.*, **41**, 407–421.

Yoshida, I., Yuasa, N., Horiuchi, T. and Tsubahara, H. (1975). *Natl Inst. Anim. Hlth Q.*, **15**, 1–7.

Zacek, D., Steward-Brown, B. and Nordgren, B. (1992). *Proceedings of the 43rd North Central Avian Disease Conference Minneapolis, Minnesota, October 4–6*, pp. 118, AAAP, Kennett Square, Pennsylvania.

Zhang, Y. and Sharma, J. M. (2001). *Avian Dis.*, **45**, 639–645.

Zhang, Y. and Sharma, J. M. (2003). *Dev. Comp. Immunol.*, **27**, 431–438.

Future strategies for controlling Marek's disease

14

ISABEL M. GIMENO
Avian Disease and Oncology Laboratory, East Lansing, Michigan, USA

Introduction

Research on Marek's disease (MD) has achieved a number of successes over the last 40 years. Chief among these has been the development of an effective system for disease control. Without this control, MD would have a devastating effect on the poultry industry, as happened in the 1960s. MD control is based on three criteria: effective vaccination, good biosecurity and selection for genetic resistance. Biosecurity and genetic resistance are currently used as important adjuncts to vaccination rather than as primary control strategies.

Despite the high efficacy of the vaccines, MD is still a major concern for the poultry industry. The current systems of control sometimes fail, and unexpected outbreaks continue to occur. In addition, currently available vaccines have not prevented the evolution of MDV to more virulent forms against which protection is difficult (Witter, 1997). Finally, because of the highly cell-associated nature of MD vaccines, the costs involved in the handling and administration of vaccines remain very high.

With better knowledge of the chicken genome and progress in the study of molecular biology of MD herpesvirus (MDV), it is likely that new approaches to MD control will be developed. Recombinant vaccines and MD-resistant transgenic chickens should provide useful tools in the future. However, new solutions are not achievable until more basic knowledge of the virus, host and MD pathogenesis is available. In addition, improving current methods of MD control will require the commitment of research institutions, vaccine manufacturers and breeding companies; the process will neither be easy nor inexpensive. Finally, any new solutions will have to be generally acceptable to the public and, in the case of genetically modified products, this may not be forthcoming.

Current MD situation and vaccination strategies

Assessing the worldwide incidence of MD is very difficult, since it is not a notifiable disease. Therefore, to obtain the most reliable information regarding the MD situation over the last ten years and current vaccination practices, an informal mail survey was conducted in July 2003 among 80 colleagues in several countries. A total of 66 responses representing 55 countries were received. Twenty countries were reported to have a higher incidence of MD (Plate 9). Five of them (Morocco, Nigeria, Portugal, Peru and Russia) have experienced economic losses due to MD since the 1990s. By contrast, 13 countries that had suffered severe losses from MD in the 1990s considered that the disease was currently under control.

Several factors were proposed that could have influenced changes in the incidence of MD. The co-existence of immunodepressive agents was reported in 10 out of the 20 countries currently suffering problems. Infectious bursal disease virus (IBDV), chicken infectious anaemia virus (CIAV) and reticuloendotheliosis virus (REV) were major factors. Co-infection with REV was particularly evident in China, where it not only caused complications with MD control but also difficulties with MD diagnosis (Cui *et al.*, 2003a). Nonetheless, the presence of immunodepressive viruses was not always associated with higher incidence of MD, since 14 other countries that are not currently suffering MD outbreaks reported MDV co-infection with immunodepressive viruses. The presence of very virulent plus (vv+) MDV was reported in 13 countries. However, the validity of this information is difficult to confirm, since current MDV pathotyping systems have not been standardized and it is not easy to correlate the classification by pathotypes established by Witter (1997) with virulence evaluated using other systems. Failure in handling and administration of vaccines was suggested as a major factor in 12 countries. Also, a reduction in genetic resistance was included as a negative factor in the control of MD in some of the answers.

Broilers were vaccinated in 27 countries, and in most cases serotype 3 (MDV-3), herpesvirus of turkeys (HVT), was the preferred vaccine. However, a dilution of the standard dose by a half or a quarter was reported in 12 countries. A few countries combined HVT with the serotype 2 (MDV-2), SB-1 strain (USA, Japan and Mexico), or with the MDV-1 strain, CVI988 (Rispens; in Italy and sometimes Japan). Almost all countries (the exceptions were India and Peru) used CVI988 to vaccinate breeders and layers. A combination of CVI988 and HVT was the most popular vaccine (37 countries). The administration of CVI988 alone or in combination with SB-1, or with HVT and SB-1, was also reported. Generally, one full dose of each vaccine strain was used. However, a double dose of each strain was used in China, Italy, Morocco, The Netherlands, Portugal and Spain. The revaccination of layers and/or breeders was a common practice in Colombia, India, Morocco, Peru, Portugal, Spain and Venezuela, and it was occasionally used in China, Germany, Italy, the Middle East, The Netherlands, Pakistan, Sudan, Thailand, Turkey and the UK. There was no consensus in the revaccination protocols used. Some of the reported protocols included vaccination *in ovo* followed by revaccination at hatch, or vaccination at hatching followed by revaccination 4–12 hours after hatching, or at 7, 18 or 21 days of age.

Use of *in ovo* vaccination for broilers was reported in Argentina, Spain, the USA, Turkey, Canada, Japan and Uruguay. In addition, this method is used to vaccinate broiler breeders in the USA, Germany, China and Canada. Vaccination at hatching by subcutaneous and/or intramuscular routes is used in all countries for vaccinating layers, and in many countries for vaccinating broilers and broiler breeders. France and Switzerland reported the vaccination of turkeys at hatching with the CVI988 strain.

MD vaccine was administered at the same time as other poultry vaccines, and a variety of antibiotics (ceftiofur, gentamicin, penicillin) was also included in 26 and 16 countries, respectively. These practices were more commonly used with broilers, but they were sometimes used with breeders and layers. IBDV and infectious bronchitis (IB) were the most common vaccines administered at hatching. However, Newcastle disease (ND), reovirus, fowlpox virus and egg drop syndrome (EDS-67) vaccines were also used.

Future risk of increasing the incidence of MD

The evidence that MD has continued to occur in vaccinated flocks in the last decade suggests it is likely that future outbreaks of MD will persist in causing problems around the world. Generally, at the time of an outbreak there is a tendency to improve biosecurity measures, properly manage vaccine delivery and control other immunodepressive viruses. In most cases, once the outbreak has cleared up the extra effort required to adequately control the disease is reduced, and eventually this can result in another outbreak. In addition, it has become common practice to administer MD vaccine at the same time as other poultry vaccines and/or antibiotics (discussed in Chapter 5). Such procedures should be carefully considered, since a negative effect of certain antibiotics on the quality of the MD vaccine has been established (Eidson *et al.*, 1973), and the interaction between MD vaccines and other vaccines has not been properly evaluated.

The continual evolution of MDV towards greater virulence (Witter, 1997) is another risk factor for the future control of MD. The effect of MDV vaccines on the progression of MDV virulence, despite their high efficacy, is well recognized (see Chapters 5 and 13). In the USA, it has been clearly shown that the virulence of MDV has always increased a few years after the introduction of a new, more efficient, vaccine onto the market (Witter, 1997). Most countries have already introduced the MDV-1 vaccine strain CVI988 (Rispens), which is considered to be the most protective vaccine currently available (Witter, 1998). It is difficult to predict if, or when, MDV will overcome immunity induced by CVI988, but based on past experience we should prepare for this eventuality.

MD vaccines have probably driven MDV to greater virulence because they do not induce sterilizing immunity. Vaccinated chickens constitute ecosystems where vaccine and virulent MDV coexist, and mutations and recombinations readily occur. Those viruses better able to escape immune responses will replicate more readily and spread to a greater extent. Solutions to this problem could be the construction of recombinant vaccines capable of producing sterilizing immunity, and the development of chickens genetically modified for resistance to MDV infection. However, these solutions cannot easily be achieved. In addition to the effect of vaccines, management practices (such as the use of diluted doses of vaccine and inefficient vaccination procedures) may also have contributed to the increase in MDV virulence.

Current tools for MD control

Vaccination

Modified live vaccines are the cornerstone of protection against MD. There is a large selection of licensed MD vaccines representing the three MDV serotypes. In general, three types of formulation, in ascending order of efficacy, are: HVT alone; a bivalent HVT plus an MDV-2 strain; and CVI988 (with or without the MDV-2 or -3 vaccines) (Witter, 1998). Recently a novel serotype 1 MDV vaccine (strain BH16) was introduced in Australia, and this seems to confer at least as much protection as the CVI988 vaccine (Karpathy *et al.*, 2002, 2003). Efforts to produce recombinant MD vaccines have been ongoing for over 20 years. Subunit vaccines expressing genes of MDV-1 in live vectors such as HVT and fowlpox virus have been constructed (Morgan *et al.*, 1992; Nazerian *et al.*, 1996); however, results have not been completely satisfactory and so far none of them have been as effective as conventional vaccines. Details of the current vaccines, as well as the recombinant vaccines that have been developed up to the present, are discussed in Chapter 13.

There are three strategies – protective synergism, revaccination and the use of adjuvants – that are commonly used to augment vaccine efficacy, but each of these needs further evaluation to validate or optimize its beneficial effects. Protective synergism from administering different vaccine serotypes together has been widely used for more than two decades (Schat *et al.*, 1982; Witter, 1982). It has great potential for improving the efficacy of each of the selected vaccine serotypes, but it is often used inappropriately and the mechanism(s) involved is not understood. A better understanding will be essential for the design of improved polyvalent vaccines. Revaccination is a common practice in many countries (see above). Despite several attempts, the value of revaccination has never been confirmed under experimental conditions (Ball and Lyman, 1977; Eleazer, 1978; R. L. Witter, I. M. Gimeno and S. M. Reddy, unpublished observations). A major problem is the lack of a challenge virus sufficiently virulent to overcome the immunity conferred by a single vaccination with the CVI988 vaccine. Another limitation is that under experimental conditions, factors that might reduce the efficiency of vaccination (i.e. immunodepressive diseases, administration of other vaccines or antibiotics with the MD vaccine, etc.) are absent. More studies need to be done to confirm and optimize the value of this practice. The use of adjuvants to enhance the immune response has been studied for several years. Acemannan, a polysaccharide, has been licensed as an adjuvant of MD vaccine in the USA since 1992, but it is not widely used. With the development of DNA vaccines, research into adjuvants has been emphasized. One approach could be to include DNA plasmids that direct the production of cytokines to enhance the cellular immune response (see Chapter 10), but the efficiency of such adjuvants in the control of MD is not yet known. Other novel adjuvants include CpG motif-containing oligodeoxynucleotides, which, when mixed with particulate antigens, result in the induction of cellular as well as humoral immune responses. However, no benefit could be detected when CpG motif-containing oligodeoxynucleotides were added to MD vaccines (R. L. Witter, East Lansing, personal communication).

Genetics

The importance of genetic resistance in the control of MD was reported several decades ago (Cole, 1968) and, indeed, was the first method used to control the disease. Detailed descriptions of the role of genetic resistance in MD are provided in Chapter 9. Primary breeding companies frequently include MD resistance as one of the criteria used for selection. Strategies to select for MD resistance have been reviewed recently (Bacon *et al.*, 2001a), and include, mainly, mass selection (i.e. breeding from survivors) and family selection utilizing pedigreed offspring as breeders. As described in Chapter 9, there are two groups of genes in the chicken that influence MD resistance: major histocompatibility complex (MHC) genes and non-MHC genes. The relevance of MHC genes in MD resistance has been amply demonstrated (Bacon *et al.*, 2001a) and is well characterized. Particularly remarkable is the influence that the MHC haplotype has on the efficacy of the response to MD vaccines, and how this influence depends on the serotype of the vaccine used (Bacon and Witter, 1992, 1994a, 1994b). Theoretically, the most appropriate vaccine could be selected based on the predominant MHC haplotype of a particular poultry strain. This issue has not been exploited in practice because of difficulties in determining the predominant MHC haplotype, especially with meat-type chickens. However, it is now technically possible to identify the predominant MHC haplotype in meat-type chickens, and as molecular techniques improve it may become economically feasible to characterize commercial lines (Li *et al.*, 1997; Goto *et al.*, 2002).

The non-MHC genes involved in MD resistance have not been as well studied, and for the most part are not fully characterized. Identification of non-MHC genes, however, might occur in the near future as the knowledge of the chicken genome is fast expanding. Taking advantage of discoveries from the study of the human genome, several novel techniques (microarrays, proteomics, RNA interference) are now being applied to investigate the chicken genome. The value of such techniques with regard to MD resistance is dealt with in more detail in other chapters (see especially Chapter 9) and in recent reviews (Bumstead, 1998; Bacon *et al.*, 2001a).

Biosecurity

Using the example of specific-pathogen-free flocks, MD can be eradicated from a farm if ideal conditions for biosecurity are followed. Unfortunately, the use of filtered and positive air pressure housing is not economically feasible for the poultry industry. However, a proper biosecurity plan should be of great help in controlling MD. There are two critical points that any biosecurity plan should include; reducing the initial levels of MDV on a farm by preventing the entry of MDV into a building, and avoiding contamination of the environment by preventing MDV from leaving that building. Measures to reduce initial levels of MDV (all-in all-out systems, single-age buildings etc.) have been widely stressed, but are not always followed. Measures to avoid contamination of the environment, however, are not always well recognized, and there are two issues that deserve more attention – air management, and the disposal of chicken carcasses and manure. Biofilters, usually designed to remove odours of air emanating from livestock facilities, could be used to decrease the level of MDV contamination in the effluent air from poultry farms (Spencer, 2001). Carcasses are usually burned or buried, but disposal of manure does not normally receive adequate attention. Composting has been shown to be a satisfactory alternative procedure to discarding both carcasses and manure (Spencer, 2001).

Developing new strategies for control

Vaccine development

An ideal MD vaccine would be highly protective, safe and stable, and able to induce sterilizing immunity, so as not to drive MDV to greater virulence. Poultry vaccines need to be inexpensive, easily administered and able to confer long-lasting protection. In addition, since the challenge from virulent MDV occurs early in life, MD vaccines must be administrated either to 1-day-old chicks or to 18- to 19-day-old embryos. MD vaccines should also be able to induce protection in the presence of maternal antibodies. It is not the aim of this chapter to provide a detailed review of each characteristic of the ideal vaccine, but rather to try to give some idea of what is lacking in the conventional vaccines and the limitations a new generation of vaccines might possess. Current

MD vaccines confer high levels of protection (close to 95 per cent), and are considered to be safe and stable when properly administered. Moreover, they are administered at hatching or *in ovo*, and induce long-lasting protection in the presence of maternal antibodies. There are two areas in which these vaccines need to improve:

1. None of the current MD vaccines induce sterilizing immunity or prevent challenge viruses from being harboured, thus risking the chances of MDV evolution. Efforts to develop sterilizing vaccines for other herpesviruses have so far been fruitless. However, some degree of success has been obtained at reducing the levels of herpes simplex virus (Klein *et al.*, 1984), cytomegalovirus (McDonald *et al.*, 1998) and Epstein-Barr virus (Tibbetts *et al.*, 2003) in long-term latency.

2. Most of the MDV vaccines are cell-associated, and have high production and storage costs. Cell-free lyophilized HVT vaccines are available, but their efficacy is much reduced in the presence of maternal antibodies against MDV (Witter and Burmester, 1979). Subunit vaccines using vectors such as fowlpox virus can also be lyophilized, but these provide less protection than conventional vaccines. DNA vaccines are still under development, but may have future potential as they can be stored at 4°C (Tischer *et al.*, 2002a).

More efficacious MD vaccines and sustainable strategies will be needed if the evolution of MDV towards greater virulence is to be halted. However, preventing such evolution will not be easy, and a search for more effective vaccines is imperative. This sets researchers an immense challenge. For instance, partially attenuated MDV-1 vaccines have provided high efficacy in the control of highly virulent MDV (Witter, 2002) – an approach that could provide a feasible strategy for improving on current vaccines. The major limitation of this approach is that although safe when used in chickens with maternal antibody to MDV, such vaccines retain residual pathogenicity in chickens lacking maternal antibodies. Attenuation of MDV by mutational insertion is another strategy to achieve a high level of protection. Strain RM1, attenuated by insertion of the REV LTR into the genome of the virulent MDV strain JM, is highly protective against vv+MDV, although it induces severe lymphoid organ atrophy in antibody-negative chickens (Witter *et al.*, 1997). If more effective, 'hotter' vaccines become necessary for controlling MD in the future, it may be necessary to review the current safety standards used for testing new vaccines (see Chapter 13).

Safety concerns about MD vaccines are currently focused on three areas: contamination with other agents, induction of gross lesions (mainly neoplastic in nature), and reversion to virulence. Immunosuppression induced by the currently available MDV vaccines is considered to be minor and transient (Witter, 2001). However, this might not be the case with the new generation of vaccines, since methods used to achieve attenuation are likely to be different. Strain RM1 is a good example of how oncogenicity and immunosuppressive characteristics can be dissociated. This strain lacks the ability to induce tumours, but causes severe lymphoid atrophy in susceptible chickens lacking maternal antibodies to MDV (Witter *et al.*, 1997). Other virulence factors should also be considered when future vaccines are evaluated – for instance, neurovirulence can be dissociated from oncogenicity (Gimeno *et al.*, 2001). However, MD vaccines have been associated with some adverse effects. MDV-2 and avian retroviruses synergistically enhance the induction of tumours (Fadly and Ewert, 1994; Aly *et al.*, 1996), and MDV-3 has been related to the induction of vitiligo (Erf *et al.*, 2001). Finally, there is a suspicion that MD vaccines could be involved in the development of the peripheral neuropathy syndrome that has often been misidentified as MD (Bacon *et al.*, 2001b). It is greatly desirable to avoid these adverse effects with a new generation of vaccines.

Subunit vaccines are considered to be more stable than the conventional vaccines. However, this is not the case for vaccines attenuated by the deletion of one gene or by insertional mutation where recombinations might occur. Examples of undesirable recombinations between mammalian herpesviruses have been reported. A lethal herpes simplex virus (HSV) recombinant resulted from the intergenomic recombination between two non-pathogenic viruses (Javier *et al.*, 1986). Also, an HSV with increased ocular and neurovirulence was obtained after inoculation of a mixture of type 1 HSVs (Brandt and Grau, 1990).

Development of transgenic chickens

Attempts to develop transgenic chickens have been made for several years, without much success. Two approaches could be followed to develop chickens more resistant to MD:

1. *Insertion of MDV genes that interfere with MDV pathogenicity.* As reported for other herpesviruses, if expressed in transgenic chickens, several MDV genes could have the potential to increase resistance to MDV. Lack of knowledge regarding the function of many MDV genes is currently the major limitation for this technique. Based on the studies conducted on other herpesviruses, three strategies might be possible – envelope blocking, antisense inhibition, and MDV-specific target cells. Transfection of the glycoprotein D (gD) gene of HSV into cultured cells (Johnson and Spear, 1989) resulted in resistance to HSV infection due to envelope blockage. In the case of MDV, however, it has been shown that gD is not essential for cell-to-cell spread (Anderson *et al.*, 1998), and other genes, such as UL32 could be more useful (Whittaker *et al.*, 1992; Lee *et al.*, 1996). Inhibition of MDV proliferation *in vitro* was observed with antisense MDV oligonucleotides (Kawamura *et al.*, 1991), but this technology has not been demonstrated *in vivo*. Finally, transgenic mice resistant to HSV have been developed by inserting a mutant form

of the gene ICP4 (Smith and DeLuca, 1992). However, several adverse effects have been attributed to the transgene.

2. *Insertion of chicken genes that increase the innate and cell-mediated immune response to MDV.* As discussed in Chapter 9, some MHC (chicken B-locus) haplotypes are important in conferring resistance to MD. Those regions responsible for resistance to MD have not yet been identified. Attempts to manipulate the expression of B-locus genes in transgenic chickens to obtain greater resistance to MD have so far been unsuccessful. Moreover, B-locus genes that confer resistance to MD may not confer resistance to other diseases (Macklin *et al.*, 2002), which could be an important problem for the poultry industry. Resistance to MD based on non-MHC genes is poorly understood, and requires further study. Once important allotypic variants of these non-MHC genes have been identified, it may be possible to use them for developing transgenic chickens. Of particular interest are those genes enhancing the immune response, such as cytokine genes, or genes related to the natural killer cell activity. The practical application of this transgenic approach will again depend on the lack of any adverse affects on susceptibility to other infections, autoimmune diseases or economic traits.

RNA interference (RNAi) has recently been used for manipulating gene expression, and might be very useful for the development of transgenic chickens. RNAi is a phenomenon in which small, double-stranded RNA molecules induce sequence-specific degradation of homologous single-stranded RNA (Zamore, 2002). Transgenic mice that expressed small interference RNA (siRNA) against hepatitis C RNA have been successfully developed (McCaffrey and Kay, 2002). The use of lentiviral vectors is another technology that might improve the current methods for developing transgenic chickens. Lentiviral vectors have recently been shown to be a very efficient means of generating transgenic mice (Lois *et al.*, 2002).

The use of transgenics for the future control of MD is still speculative, and does not appear to be a solution for the near future. Development of transgenic chickens, however, will be very useful for gaining a better understanding of MD pathogenesis and the mechanisms of resistance, as well as for designing new strategies for control.

What research is needed for the rational development of new strategies of control?

Two conditions are needed for the rational development of new strategies for control: basic knowledge of the MDV-host system to identify target genes, and adequate techniques to manipulate MDV and chicken genomes. Manipulation of the MDV genome has become easier with the use of BAC clones (Petherbridge *et al.*, 2002) and overlapping cosmid clones (Reddy *et al.*, 2002).

However, more knowledge about the basis for vaccinal immunity, MDV gene function and molecular pathogenesis of MDV is required.

Much work has been conducted to elucidate the protective nature of MD vaccines. As described in Chapters 10 and 13, vaccination stimulates a variety of cellular and humoral immune responses primarily directed against viral antigens. A number of MDV antigens can elicit immune responses *in vitro* (Schat and Markowski-Grimsrud, 2001), but it is not yet clear which viral antigens are important for eliciting immune responses *in vivo*. Activation of T cells and the critical role that this plays in MD pathogenesis, as well as in the immune response, require further study. Activated T cells are target cells for latency and transformation, as well as being the effector cells of the immune system. Furthermore, it has been shown that ability to induce early activation of T cells is a common characteristic of the most protective vaccines (Gimeno *et al.*, 2004).

One of the more critical goals of current MD research is elucidation of MDV gene function. This information is vital not only for identifying the basis of virulence, but also for better understanding vaccine-induced responses and relevant events in MDV pathogenesis. The most widely used strategy to study gene function relies on the use of mutant MDV in which one or more genes have been inactivated or modified. Results obtained so far suggest that MDV virulence is very complex, indicating the limitations of this approach. Deletion of crucial genes for pathogenicity does not always cause complete loss of oncogenicity (Silva *et al.*, 2002; Cui *et al.*, 2003b; Gimeno *et al.*, 2003). Also, MDV genes are usually involved in more than one biological process, and deletion of some genes relevant to virulence can result in a severe reduction of replication and antigenicity (Cui *et al.*, 2003b; Gimeno *et al.*, 2003). Complementary information about gene expression *in vivo* and interaction between proteins could be generated by the use of microarrays and proteomics, once these technologies are standardized for chickens. Another promising technology to elucidate gene function is the use of RNAi (see the section on transgenics). This strategy needs to be optimized, but could prove very useful in the future control of MD (Wang *et al.*, 2003).

Understanding the molecular pathogenesis of MDV will ultimately assist in the control of MD. There are three major areas for study: molecular mechanisms involved in latency (recently reviewed by Morgan *et al.*, 2001); molecular mechanisms involved in oncogenicity (see Chapter 4); and mechanisms by which MDV infects cells, replicates in the feather-follicle epithelium and transmits (see Chapter 6). Very little is known about infection of cells through cell-free virus, and putative receptors have yet to be identified. Glycoproteins are presumed to be important for infection of cells and transfer of virus from cell to cell. Glycoprotein B (gB) induces the production of neutralizing antibodies, which suggests that gB might be relevant for cell attachment and/or penetration (Ikuta *et al.*, 1984). A gC deleted mutant resulted in an attenuated phenotype with decreased infectivity, horizontal transmission and oncogenicity, although a revertant virus needs to be constructed to confirm these results (Morgan *et al.*, 1996). Unlike with other herpesviruses, gD is not essential for

MDV transmission (Anderson *et al.*, 1998). Mutant viruses with deleted gM, gI or gE were unable to transfer infectivity from infected to uninfected cells (Schumacher *et al.*, 2001; Tischer *et al.*, 2002b). A better understanding of the MDV genes involved in MDV infection and transmission is needed, and will be crucial for developing sterilizing vaccines and transgenic chickens resistant to MDV infection.

Summary

The poultry industry cannot rely solely on the use of effective vaccines to control MD, and biosecurity and genetics will always be important adjuncts to vaccination. In the future, improved control of MD will require two approaches. The first consists of better application of the tools that are currently available. Proper measures of biosecurity, adequate vaccination practices and good control of other immunosuppressive viruses (CIAV, IBDV, REV) require further attention. The second approach is the development of improved tools that prevent the MDV evolving to greater virulence (sterilizing vaccines or chickens resistant to MDV infection), reducing the economic costs associated with MD vaccination (efficient cell-free vaccines) or conferring greater protection against the most virulent pathotypes of MDV (see Figure 14.1).

Effective control of MD will require some compromise and efforts at every level in the poultry production sector (Figure 14.2). Research institutions (academic, government, research and development departments of companies)

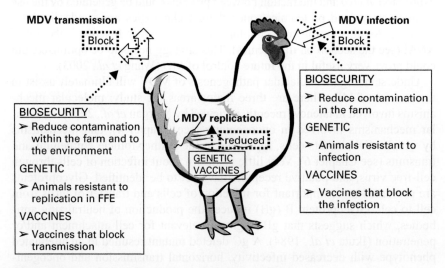

Figure 14.1 Critical points in the epidemiology of MD and possible strategies of control. Those measures that are currently available are underlined.

need to obtain the necessary basic knowledge, new technologies and strategies. Breeding companies need to develop methods for selecting chickens with innate resistance and better responsiveness to vaccines. Moreover, breeding companies could have a major impact by providing better storage facilities and proper administration of MD vaccines, as well as by advising the producers about adequate management practices. Vaccine manufacturers will be responsible for developing suitable procedures to generate the next generation of improved vaccines.

The future prospects for the control of MD appear to be fairly optimistic. We know what needs to be done, and many of the technologies are already developed. However, the reality does not match this optimism. Basic research suffers from limited funding to study MD, as other diseases are currently considered a greater priority for veterinary research. The level of MD is not significant in many countries, and thus current research efforts are limited. Due to the low costs of disease losses, it might not be economically feasible for companies to increase investment in husbandry or vaccines. Moreover, profit margins on vaccines are too small to encourage further research efforts, and consequently vaccine manufacturers are unlikely to invest more in the development of new MD vaccines. Finally, some of the resulting products, such as transgenic chickens, may not be acceptable to the general public.

It therefore seems that future measures for the control of MD will be driven by the evolution of the virus itself. If MD continues to remain at a low level, there is likely to be a gradual decline in research effort and investment, making it highly unlikely that new methods of control will be developed. In this case,

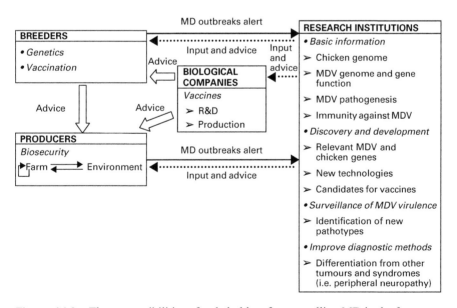

Figure 14.2 The responsibilities of stakeholders for controlling MD in the future.

we will already have reached the highest level of control for MD that can be achieved, if current tools are used effectively. If MDV continues to evolve to greater virulence, the development of new methods for control, as described above, will be imperative. We can only hope that, should this happen, sufficient expertise will be available to respond to the situation.

Acknowledgements

I thank Drs Larry Bacon, Bruce Calnek, Robert Silva, Lloyd Spencer and Richard Witter for sharing their thoughts on the future of Marek's disease. Also, I want to thank all my colleagues and friends in the poultry industry for the information that they have provided.

References

Aly, M. M., Witter, R. L. and Fadly, A. M. (1996). *Avian Pathol.*, **25**, 81–94.
Anderson, A. S., Parcells, M. S. and Morgan, R. W. (1998). *J. Virol.*, **72**, 2548–2553.
Bacon, L. D. and Witter, R. L. (1992). *Avian Dis.*, **36**, 378–385.
Bacon, L. D. and Witter, R. L. (1994a). *Poultry Sci.*, **73**, 481–487.
Bacon, L. D. and Witter, R. L. (1994b). *Avian Dis.*, **38**, 65–71.
Bacon, L. D., Hunt, H. D. and Cheng, H. H. (2001a). In *Current Topics in Microbiology and Immunology* (ed. K. Hirai), pp. 121–142. Springer-Verlag, Berlin.
Bacon, L. D., Witter, R. L. and Silva, R. F. (2001b). *Avian Pathol.*, **30**, 487–499.
Ball, R. F. and Lyman, J. (1977). *Avian Dis.*, **21**, 440–444.
Brandt, C. R. and Grau, D. R. (1990). *Invest. Ophthal. Vis. Sci.*, **31**, 2214–2223.
Bumstead, N. (1998). *Avian Path.*, **27**, S78–S81.
Cole, R. K. (1968). *Avian Dis.*, **12**, 9–28.
Cui, Z., Zhang, Z., Jiang, S. and Zhou, J. (2003a). In *XIII Congress of the World Veterinary Poultry Association*, Denver, pp. 144, AAAP, Kennett Square, Pennsylvania.
Cui, X., Lee, L. F. and Reddy, S. M. (2003b). In *XIII Congress of the World Veterinary Poultry Association,* Denver, pp. 143, AAAP, Kennett Square, Pennsylvania.
Eidson, C. S., Kleven, S. H. and Anderson, D. P. (1973). *Poultry Sci.*, **52**, 755–760.
Eleazer, T. H. (1978). *Poultry Digest*, **37**, 154.
Erf, G. F., Bersi, T. K., Wang, X. L. *et al.* (2001). *Pigment Cell Res.*, **14**, 40–46.
Fadly, A. M. and Ewert, D. L. (1994). In *Interactions Between Retroviruses and Herpesviruses* (eds H. J. Kung and C. Wood), pp. 1–9. World Scientific Publishing Co., New Jersey.
Gimeno, I. M., Witter, R. L., Hunt, H. D. *et al.* (2001). *Avian Pathol.*, **30**, 397–409.
Gimeno, I. M., Hunt, H. D., Witter, R. L. *et al.* (2003). In *XIII Congress of the World Veterinary Poultry Association*, Denver, pp. 101, AAAP, Kennett Square, Pennsylvania.
Gimeno, I. M., Witter, R. L., Hunt, H. D. *et al.* (2004). *Avian Pathol.*, **33**, 59–68.
Goto, R. M., Afanassieff, M., Ha, J. *et al.* (2002). *Poultry Sci.*, **81**, 1832–1841.
Ikuta, K., Ueda, S., Kato, S. and Hirai, K. (1984). *Microbiol. Immunol.*, **28**, 923–933.
Javier, R. T., Sedarati, F. and Stevens, J. G. (1986). *Science*, **234**, 746.
Johnson, R. M. and Spear, P. G. (1989). *J. Virol.*, **63**, 819–827.
Karpathy, R. C., Firth, G. A. and Tannock, G. A. (2002). *Aust. Vet. J.*, **80**, 39–44.

Karpathy, R. C., Firth, G. A. and Tannock, G. A. (2003). *Aust. Vet. J.*, **81**, 222–225.

Kawamura, M., Hayashi, M., Furuichi, T. *et al.* (1991). *J. Gen. Virol.*, **72**, 1105–1111.

Klein, R. J., Kaley, L. A. and Friedman-Kien, A. E. (1984). *Vaccine*, **2**, 219–223.

Lee, L. F., Wu, P. and Sui, D. (1996). In *Current Research on Marek's Disease* (eds R. F. Silva, H. H. Cheng, P. M. Coussens *et al.*), pp. 245–250. American Association of Avian Pathologists, Kennett Square, Pennsylvania.

Li, L., Johnson, L. W. and Ewald, S. J. (1997). *Anim. Gen.*, **28**, 258–267.

Lois, C., Hong, E. J., Pease, S. *et al.* (2002). *Science*, **295**, 868–872.

Macklin, K. S., Ewald, S. J. and Norton, R. A. (2002). *Avian Pathol.*, **31**, 371–376.

McCaffrey, A. P. and Kay, M. A. (2002). *Gene Therapy*, **9**, 1563.

McDonald, M. R., Li, X. Y., Stenberg, R. M. *et al. J. Virol.*, **72**, 442–451.

Morgan, R. W., Gelb, J. Jr., Schreurs, C. S. *et al.* (1992). *Avian Dis.*, **36**, 858–870.

Morgan, R. W., Anderson, A., Kent, J. and Parcells, M. S. (1996). In *Current Research on Marek's Disease* (eds R. F. Silva, H. H. Cheng, P. M. Coussens, *et al.*), pp. 207–212. American Association of Avian Pathologists, Kennett Square, Pennsylvania.

Morgan, R. W., Xie, Q., Cantello, J. L. *et al.* (2001). In *Current Topics in Microbiology and Immunology* (ed. K. Hirai), pp. 223–244. Springer-Verlag, Berlin.

Nazerian, K., Witter, R. L., Lee, L. F. and Yanagida, N. (1996). *Avian Dis.*, **40**, 368–376.

Petherbridge, L., Howes, K., Baigent, S. J. *et al.* (2002). *Workshop on Molecular Pathogenesis of Marek's Disease and Avian Immunology, Limassol, Cyprus*, p. 47, Kimron Institute, Beit Dagan, Israel.

Reddy, S. M., Lupiani, B., Gimeno, I. M. *et al.* (2002). *Proc. Natl. Acad. Sci. USA*, **99**, 7054–7059.

Schat, K. A. and Markowski-Grimsrud, C. J. (2001). In *Current Topics in Microbiology and Immunology* (ed. K. Hirai), pp. 91–120. Springer-Verlag, Berlin.

Schat, K. A., Calnek, B. W. and Fabricant, J. (1982). *Avian Pathol.*, **11**, 593–605.

Schumacher, D., Tischer, B. K., Reddy, S. M. and Osterrieder, N. (2001). *J. Virol.*, **75**, 11307–11318.

Silva, R. F., Lupiani, B. and Reddy, S. M. (2002). *Workshop on Molecular Pathogenesis of Marek's Disease and Avian Immunology, Limassol, Cyprus*, p. 49, Kimron Institute, Beit Dagan, Israel.

Smith, C. A. and DeLuca, N. A. (1992). *Virology*, **191**, 581–588.

Spencer, J. L. (2001). In *XII International Congress of the World's Poultry Science Association*, Cairo, pp. 1–8, WPSA, Cairo.

Tibbetts, S. A., McClellan, J. S., Gangappa, S. *et al.* (2003). *J. Virol.*, **77**, 2522–2529.

Tischer, B. K., Schumacher, D., Beer, M. *et al.* (2002a). *Workshop on Molecular Pathogenesis of Marek's Disease and Avian Immunology, Limassol, Cyprus*, p. 46, Kimron Institute, Beit Dagan, Israel.

Tischer, B. K., Schumacher, D., Messerle, M. *et al.* (2002b). *J. Gen. Virol.*, **83**, 997–1003.

Wang, Q. C., Nie, Q. H. and Feng, Z. H. (2003). *World J. Gastr.*, **9**, 1657–1661.

Whittaker, G. R., Bonass, W. A., Elton, D. M. *et al.* (1992). *J. Gen. Virol.*, **73**, 2933–2940.

Witter, R. L. (1982). *Avian Pathol.*, **11**, 49–62.

Witter, R. L. (1997). *Avian Dis.*, **41**, 149–163.

Witter, R. L. (1998). *Poultry Sci.*, **77**, 1197–1203.

Witter, R. L. (2001). In *Current Topics in Microbiology and Immunology* (ed. K. Hirai), pp. 58–90. Springer-Verlag, Berlin.

Witter, R. L. (2002). *Avian Dis.*, **46**, 925–937.

Witter, R. L. and Burmester, B. R. (1979). *Avian Pathol.*, **8**, 145–156.

Witter, R. L., Li, D., Jones, D. *et al.* (1997). *Avian Dis.*, **41**, 407–421.

Zamore, P. D. (2002). *Science*, **296**, 1265–1269.

Conclusions

15

FRED DAVISON, VENUGOPAL NAIR and PAUL-PIERRE PASTORET

Institute for Animal Health, Compton Laboratory, Newbury, Berkshire, UK

It is almost 100 years since József Marek, in 1907, first described the form of polyneuritis in four cockerels – the disease that later came to bear his name. It was another 60 years before the causative agent, Marek's disease herpesvirus (MDV), was identified. However, once identified, it was only a very short time until the first live attenuated vaccine was developed. Since then, despite the widespread use of such vaccines, Marek's disease (MD) has continued to be a concern for the poultry industry. In this book, we have tried to highlight some of the scientific progress that has been made in our understanding of the various factors contributing to this complex, some would say fascinating, disease and its causative herpesvirus.

The different contributors to this book have shown just how the subject of MD impinges on so many biological disciplines, involving both the virus and its host. They have made it clear that current problems in controlling this insidious disease can no longer rely on a 'trial and error' approach, but need to be tackled in a rational and coherent way. It seems self-evident that control of MD will only be possible by improving our knowledge base and bringing together experts from different fields to devise new strategies for control – strategies that are sustainable and long-term (see Chapter 14).

Marek's disease is clearly a problem that came to the fore through the development of modern practices in poultry farming. Over the years, the success of poultry industries around the world has resulted in eggs and poultry meat becoming important sources of 'healthy' dietary proteins in most developed countries, and essential sources of protein in many developing countries. Ironically, without the growth and intensification of modern chicken egg and meat production systems, MD would probably have remained only a minor problem afflicting domesticated chickens. As pointed out in Chapter 2, what we now know as MD was originally described as a neurological condition causing paralysis; in the 1920s, visceral lymphoma only arose in 10 per cent of the cases of fowl paralysis. However, as poultry production increased this lymphomatous form of the disease came to predominate and, just before the advent of effective MD vaccines, MD was devastating poultry industries around the world. The introduction of vaccination was initially a huge success and led to the effective control of MD (discussed in Chapter 13), but within 10 years outbreaks of MD were being reported and hypervirulent pathotypes of MDV were being isolated from vaccinated poultry. The use of more aggressive vaccination – 'hotter' vaccines, such as CVI988 (the Rispens vaccine) and bivalent and trivalent vaccines – is generally agreed to have driven MDV to evolve to even greater levels of virulence. In some areas only the serotype 1 CVI988 vaccine, either alone or in combination with other serotypes, is now an effective measure. However, CVI988, perhaps the most efficacious MD vaccine so far, is derived from a serotype 1 MDV that is weakly oncogenic in susceptible genotypes of chickens (discussed in Chapter 13). This has led to the important question (Witter, 1997): where do we go if hypervirulent MDV pathotypes evolve that can break through the protection of CVI988?

In recent years there have been a number of major technological achievements concerning the family of viruses that belong to the different serotypes of MDV. Because of their close relationship, the members have been placed in the taxonomic family *Mardivirus*, and the entire genomes of representatives of each of the three serotypes have been sequenced. In Chapter 3, Osterrieder and Vautherot conclude that this sequence information indicates that the three different MDV serotypes represent three distinct, individual species that have undergone parallel evolution. They propose the revolutionary step of changing the *Mardivirus* nomenclature to reflect this situation, with only serotype 1 virus strains retaining the name 'Marek's disease virus'. Time will tell if this recommendation becomes generally accepted by those with an intimate knowledge of the *Mardivirus* family. Further practical advances have come with the cloning and mutagenesis of the whole MDV genome into a bacterial artificial chromosome (BAC) library, and successful MDV reconstitution in both cell culture and the chicken. This is a major step forward, providing an opportunity to delete or add genes to the MDV genome, and allowing the construction of novel recombinant viruses and vaccines. It should now be feasible systematically to investigate the roles of different genes in MDV pathogenesis and oncogenesis. For instance, this opens up the possibility of identifying regulatory genes that are essential and cannot easily be modified without preventing MDV replication. If such genes or their products are (or can be made) targets for the host's immune responses, they could provide a useful means for inducing sterilizing immunity to MDV. They could be incorporated into novel, more efficacious MD vaccines. Compared to many DNA tumour viruses, the onset of MDV-induced tumours is very rapid, suggesting a direct involvement of virus-encoded genes in the transformation of lymphocytes. Among these, the MDV-encoded Meq gene shows all the hallmarks of a potent oncogene. In Chapter 4, Kung and Nair provide a detailed account of recent developments in identifying the molecular pathways of Meq, particularly the involvement of host genes Jun and Ski; these give further insights into Meq's complex functions. MD tumour cells express high levels of Meq, which is correlated with the expression of the chicken homologue of CD30, the Hodgkin's antigen (see Chapter 10). These findings should surely increase our interest in MD as a natural model for studying the molecular mechanisms of oncogenesis.

With the introduction of new technologies in genetics and immunology, we have begun to gain a fundamental understanding of the importance of the host's innate and acquired responses to MDV, and how these can produce life-long MDV latency or lead to transformation of T cells, tumour formation and/or paralysis and death. Clearly there is still a long way to go before useful host characteristics can be effectively harnessed to improve MD resistance. The seminal work of Witter (1997) has highlighted the problems of increasing MDV virulence; however, until now the strategy to control increasingly virulent pathotypes has been to introduce more aggressive vaccine regimes (see Chapters 13 and 14). The risk is that 'hotter' vaccines will border on being harmful, themselves causing immunosuppression or tumours in susceptible chicken genotypes. Recently, mathematical modelling has suggested that vaccines directed against pathogen replication tend to drive that pathogen to increasing virulence (Gandon *et al.*, 2001). Current MD vaccines are mainly directed at controlling MDV replication, and Gandon *et al.* (2001) cite MDV as the prime example of this phenomenon. In contrast, vaccines directed against infection (rather than replication) were predicted to have a neutral, or even negative, effect on pathogen virulence. Clearly there are lessons here for the future development of MD vaccines. We know very little of the early stages of natural infection via the respiratory tract (see Chapter 6). Most studies on host responses have ignored the possibility of barriers to natural infection by MDV and early host responses before the virus replicates in the lymphoid tissues. It is conceivable that current vaccines, although good at controlling cytolytic infection, are relatively poor at preventing uptake and initial events in infection. Better models that mimic natural MDV challenge could prove valuable for identifying early protective responses that could be exploited to improve vaccinal protection. Another stage in the MDV lifecycle that has received little attention is the production of cell-free MDV and the shedding of infectious virus from the feather-follicle epithelium (FFE). It is generally considered that once infected with MDV, chickens shed virus throughout their lives, even if they were vaccinated as chicks (see Chapters 6 and 14). As a consequence, infected chickens are a continual source of infectious virus. Despite the importance of fully productive MDV infection to the persistence of the virus in a flock, there have been relatively few studies on the subject, although lymphocyte infiltration and lesions in the FFE and feather pulp have been documented (see Chapter 7). It would useful to know something about the immune responses at the sites of fully productive MDV so that, perhaps, vaccine responses could be targeted here and immunity used to break the cycle of infection.

In Chapters 8 and 10 it was pointed out that the types of immune responses invoked by MDV are probably critical for determining the outcome of infection in chickens of different genotypes. This is not a new concept, since it has been known for some time that the polarization of the immune response to either T helper-1 or T helper-2 responses can make a difference between survival or death when some genotypes of mice are exposed to a particular pathogen. Cytokines drive and direct the immune responses having a critical effect on the course of the infection and chemokines attract specific populations of effector cells to the sites of infection (see Chapter 6). With developments in chicken genomics and cloning technology the arsenal of chicken cytokines and chemokines is rapidly expanding, allowing their use to identify protective immune responses. They could also be exploited as natural adjuvants. By incorporating such adjuvants into recombinant vaccines, it should be possible to drive and direct better protective immune responses, so improving immunity upon later challenge from hypervirulent MDV.

Before MD vaccines became available there was a great deal of interest in genetic resistance to MD; however, with the introduction of vaccines, interest in harnessing genetic resistance in breeding programmes has waned. The growing evidence of increased virulence of MDV has led to a revival of interest in genetic resistance. Fortunately there has continued to be a small group of researchers investigating the genetic basis of resistance to MD – resistance associated with genes of the major histocompatibility complex (MHC) and that associated with genes outside of the MHC, such as the MDV1 region. It seems highly likely that the MDV1 region could be the homologue of the mammalian natural killer complex that has been associated with resistance to other herpesviruses (see Chapter 9). The interaction(s) between genetic resistance associated with genes within the MHC and outwith the MHC merits investigation, especially with regard to the influences on immune responses. With the recent availability of the sequence of the chicken genome, our ability to identify protective genes should improve enormously. The ability to harness such genes in breeding programmes will be very valuable. Newer methods of modifying the genome should markedly increase the efficiency of transgenic technology, which could be harnessed to generate transgenic chickens with increased MD resistance. Recently, the revolutionary new technology of RNA interference (RNAi), delivered through highly efficient lentiviral vectors, has been used successfully to generate transgenic mice. The use of similar approaches to generate pathogen-resistant chickens using RNAi of important viral genes could be a very attractive and novel method for controlling MD in the future. Whether the products of this technology will become accepted by the general public remains to be seen.

Study of MD and its causative herpesvirus, and the development of MD vaccines, have provided many important insights into fundamental mechanisms of disease control and led to a number of important firsts – the first widely used anti-cancer vaccine, and the use of *in ovo* vaccination technology. Most importantly, the problems of controlling MD by vaccination teach us that intensive use of vaccines may have a malign influence on the evolution of the pathogen itself. With new technological developments, our ability to gain a fundamental understanding of herpesvirology, tumorigenesis, immunology and the genetic basis of disease resistance should be greatly improved. Investigations on MD also have the advantage that the pathogen can be studied in its natural host at the molecular, cellular, animal and population levels. The lessons to be learnt are likely to be relevant not only to poultry and other agricultural species, but also to humans.

References

Gandon, S., Mackinnon, M. J., Nee, S. and Read, A. F. (2001). *Nature*, **414**, 751–755.
Witter, R. L. (1997). *Avian Dis.*, **41**, 149–163.

Appendix: Key references

Biggs, P. M. (1961). A discussion on the classification of the avian leucosis complex and fowl paralysis. *Br. Vet. J.*, **117,** 326–334.

Biggs, P. M. and Payne, L. N. (1964). The relationship of Marek's disease (neurolymphomatosis) to lymphoid leukosis. *Natl Cancer Inst. Monogr.*, **17,** 83–97.

Biggs, P. M. and Payne, L. N. (1967). Studies on Marek's disease. I. Experimental transmission. *J. Natl Cancer Inst.*, **39,** 267–280.

Biggs, P. M., Churchill, A. E., Rootes, D. G. and Chubb, R. C. (1968). The etiology of Marek's disease – an oncogenic herpes-type virus. Virus-induced immunopathology. *Persp. Virol.*, **6,** 211–230.

Buckmaster, A. E., Scott, S. D., Sanderson, M. J. *et al.* (1988). Gene sequence and mapping data from Marek's disease virus and herpesvirus of turkeys: implication for herpesvirus classification. *J. Gen. Virol.*, **69,** 2033–2042.

Calnek, B. W. (1986). Marek's disease – a model for herpesvirus oncology. *CRC Crit. Rev. Microbiol.*, **12,** 293–320.

Calnek, B. W., Adldinger, H. K. and Kahn, D. E. (1970). Feather follicle epithelium: a source of enveloped and infectious cell-free herpesvirus from Marek's disease. *Avian Dis.*, **14,** 219–233.

Cauchy, L. (1990). Tumeurs du système immunitaire dues à des virus herpes. In *Immunologie animale* (eds P.-P. Pastoret, A.Govaerts and H. Bazin), pp.377–381. Médecine-Sciences, Flammarion.

Churchill, A. E. and Biggs, P. M. (1967). Agent of Marek's disease in tissue culture. *Nature*, **215,** 528–530.

Churchill, A. E. and Biggs, P. M. (1968). Herpes-type virus isolated in cell culture from tumours of chickens with Marek's disease. II. Studies *in vivo*. *J. Natl Cancer Inst.*, **41,** 951–956.

Churchill, A. E., Payne, L. N. and Chubb, R. C. (1969). Immunisation against Marek's disease using a live attenuated virus. *Nature*, **221,** 744–747.

Delecluse, H. J., Schüller, S. and Hammerschmidt, W. (1993) Latent Marek's disease virus can be activated from its chromosomally integrated state in herpesvirus-transformed lymphoma cells. *EMBO J.*, **12,** 3277–3286.

Hirai, K. (2001). *Marek's Disease*. Springer-Verlag, Berlin.

Kawamura, H., King, D. J. Jnr and Anderson, D. P. (1969). A herpesvirus isolated from kidney cell culture of normal turkeys. *Avian Dis.*, **13,** 853–863.

Kovács, F., Kvörös, K. and Solti, L. (2002). 50 éve halt meg Marek József. *Magyar Állatorvosok Lapja*, **12**(124), 707–718.

Lee, L. F., Wu, P. and Sui, D. *et al.* (2000). The complete unique long sequence and overall genomic organization of the GA strain of Marek's disease virus. *Proc. Natl Acad. Sci. USA*, **97**, 6091–6096.

Marek, J. (1907). Polyneuritis Kakasokban. *Állatorvosok Lapja*, 29 June.

Marek, J. (1907). Multiple Nervenenzundung (Polyneuritis) bei Hühnern. *Dtsch Tierärztl. Wochenschr.*, **15**, 417–421.

Marek, J. (1952). In memoriam. *Magyar Állatorvosok Lapja*, Szepiember, 7 Évfolyam–9 Szám, pp. 257–259.

Milan, M. (1998). *Prof. MVDr. Et PhDr. Jozef Marek, Veterinár Svetového mena. K 130 výrociu narodenia*. Obecn" úrad Horná Streda Marec.

Mocsy, J. (1952). Marek József, a tudós, tanár és ember. *Magyar Állatorvosok Lapja*, November, 7, Évpolyam–11 Szám.

Nazerian, K., Solomon, J. J., Witter, R. L. and Burmester, B. R. (1968). Studies on the etiology of Marek's disease. II. Finding a herpesvirus in cell culture. *Proc. Soc. Exp. Biol. Med.* **127**, 177–182.

Okazaki, W., Purchase, H. G. and Burmester, B. R. (1970). Protection against Marek's disease by vaccination with a herpesvirus of turkeys. *Avian Dis.*, **14**, 413–429.

Pasteur, L. (1880). De l'atténuation du virus du choléra des poules. *Comptes rendus de l'Académie des Sciences, Paris*, **91**, 673–680.

Pastoret, P.-P. (1997). Other models of relevance to EBV. In *IARC Monographs on the Evaluation of Carcinogenic Risks to Humans, Vol. 70, Epstein-Barr virus and Kaposi's Sarcoma, herpesvirus/human herpesvirus 8*, pp. 212–213. World Health Organisation – International Agency for Research on Cancer, Lyon.

Payne, L. N. (1973). Marek's disease: A possible model for herpesvirus-induced neoplasm in man. In *Proceedings of the 3rd International Symposium of the Princess Takamatsu Cancer Research Fund on Analytic and Experimental Epidemiology of Cancer* (ed. W. Nakahara, T. Hirayama, K. Nishiska and H. Sugano), pp. 235–257. University of Tokyo Press, Tokyo.

Payne, L. N. (1979). Marek's disease in comparative medicine. Editorial. *J. R. Soc. Med.*, **27**, 635–638.

Payne, L. N. (1985). *Marek's Disease: Scientific Basis and Practice*. Martinus Nijhoff, Boston.

Sharma, J. M. (1998). In *Avian Immunology: Handbook of Vertebrate Immunology* (ed. P.-P. Pastoret, P. Griebel, H. Bazin, and A. Govaerts), pp. 73–136. Academic Press, San Diego.

Solomon, J. J., Witter, R. L., Nazerian, K. and Burmester, B. R. (1968). Studies on the etiology of Marek's disease. I. Propagation of the agent in cell culture. *Proc. Soc. Exp. Biol. Med.*, **127**, 173–177.

Tulman, E. R., Afonso, C. L. and Lu, Z. *et al.* (2000). The genome of a very virulent Marek's disease virus. *J. Virol.*, **74**, 7980–7989.

Witter, R. L., Nazerian, K., Purchase, H. G. and Burgoyne, G. H. (1970). Isolation from turkeys of a cell-associated herpesvirus antigenically related to Marek's disease virus. *Am. J. Vet. Res.*, **31**, 525–538.

Index

Printed and bound by CPI Group (UK) Ltd, Croydon, CR0 4YY

08/05/2025

01864881-0001